MW00844581

SAFE WORK PRACTICES FOR
WASTEWATER TREATMENT PLANTS

HOW TO ORDER THIS BOOK

BY PHONE: 800-233-9936 or 717-291-5609, 8AM–5PM Eastern Time

BY FAX: 717-295-4538

BY MAIL: Order Department
Technomic Publishing Company, Inc.
851 New Holland Avenue, Box 3535
Lancaster, PA 17604, U.S.A.

BY CREDIT CARD: American Express, VISA, MasterCard

BY WWW SITE: http://www.techpub.com

PERMISSION TO PHOTOCOPY–POLICY STATEMENT

Authorization to photocopy items for internal or personal use, or the internal or personal use of specific clients, is granted by Technomic Publishing Co., Inc. provided that the base fee of US $3.00 per copy, plus US $ .25 per page is paid directly to Copyright Clearance Center, 222 Rosewood Drive, Danvers, MA 01923, USA. For those organizations that have been granted a photocopy license by CCC, a separate system of payment has been arranged. The fee code for users of the Transactional Reporting Service is 1-56676/96 $5.00 + $ .25.

SAFE WORK PRACTICES FOR WASTEWATER TREATMENT PLANTS

FRANK R. SPELLMAN, Ph.D., CSP

Environmental Health & Safety Manager
Hampton Roads Sanitation District
Virginia Beach, Virginia

TECHNOMIC
PUBLISHING CO., INC.

LANCASTER · BASEL

Safe Work Practices for Wastewater Treatment Plants

a **TECHNOMIC**®publication

Published in the Western Hemisphere by
Technomic Publishing Company, Inc.
851 New Holland Avenue, Box 3535
Lancaster, Pennsylvania 17604 U.S.A.

Distributed in the Rest of the World by
Technomic Publishing AG
Missionsstrasse 44
CH-4055 Basel, Switzerland

Copyright © 1996 by Technomic Publishing Company, Inc.
All rights reserved

No part of this publication may be reproduced, stored in a
retrieval system, or transmitted, in any form or by any means,
electronic, mechanical, photocopying, recording, or otherwise,
without the prior written permission of the publisher.

Printed in the United States of America
10 9 8 7 6 5 4 3 2

Main entry under title:
  Safe Work Practices for Wastewater Treatment Plants

A Technomic Publishing Company book
Bibliography: p. 219
Includes index p. 221

Library of Congress Catalog Card No. 95-62354
ISBN No. 1-56676-406-8

For
Jennifer Overman Rodrigues,
the consummate safety associate

Safety practitioners in wastewater treatment who have had experience with occupational health and safety problems in the field have pointed to the need for a book that is comprehensive in scope and directly applicable to conditions actually encountered in practice. Additionally, there has been a call for a safety text that is user friendly.

Several standard texts adequately cover the specialized aspects of occupational health and safety, but little detailed information is available in one volume dealing specifically with the wastewater industry.

In this text emphasis is placed on the practical applications of occupational health, safety, and safe work practices to hazard control in wastewater treatment and collection. A special effort has been made to include lessons learned as they relate to safety. Examples are used freely to help in the implementation and use of the subject matter.

Since the safety field is extremely broad, the following subjects are specifically covered in this text:

(1) Safety programs
(2) Training requirements
(3) Documentation
(4) Safety auditing
(5) Safety equipment required for wastewater work
(6) Safe work practices

All topic areas are presented in an easy-to-read and easy-to-understand format.

I am indebted to many individuals for their technical review and suggestions for this text. However, the author must, of course, bear full responsibility for the content and errors, if any.

Special thanks are due Jennifer Overman Rodrigues, Industrial Hygienist, Hampton Roads Sanitation District (HRSD), for her input, review of materials, and never-ending inspiration.

I am grateful to James R. Borberg, P.E., General Manager HRSD, and Commission Members for allowing me free access to and use of HRSD safety materials that are included in this text. Additionally, I am grateful to HRSD employees in all departments who comprised a special project safety team that reviewed the safe work practices I wrote for use at HRSD. This safety project team was led by Mardane McLemore, P.E., and consisted of team members: Robert Rutherford, Darlene Raiford, Bob Norris, Richard Roberts, David Morse, Pete Henderson, Rick Baumler, Larry Barr, Guy Aydlett, and Sam Reaves. Without the safety project team's critical analysis and constructive criticism, this text would not be possible.

HRSD employees contributed to this text by setting an example of safety excellence that I feel is without parallel – anywhere. HRSD's safety program has been a success for one reason: HRSD's employees.

I am especially grateful to my family for their extreme patience and for putting up with the weekends/holidays lost and the early morning departures from a warm bed to the computer terminal that was constantly calling me. Without their support this text would not be possible.

CHAPTER 1

Introduction

Several statistical reports have related historical evidence showing that the wastewater treatment industry is an extremely unsafe occupational field. This less than stellar safety performance has continued to deteriorate even in the age of the Occupational Safety and Health Act (OSH Act). As a case in point, consider the records from 1976–1977: The accident frequency rate during this period for wastewater treatment plants was 38.0, which was nearly four times the rate of 10 for the average U.S. industry. For the same period, the accident frequency rate for the worst industry reporting to the National Safety Council (N.S.C.) was 40.

The question is why is the wastewater treatment industry on-the-job injury rate so high? Several reasons help to explain this high injury rate. First, all of the major classifications of hazards exist at a wastewater treatment plant (exception radioactivity):

- oxygen deficiency
- physical injuries
- toxic gases and vapors
- infections
- fire
- explosion
- electrocution

Along with all the major classifications of hazards, other factors cause the high incidence of injury in the wastewater industry. Some of these can be attributed to the following:

- complex treatment systems
- shift work

1

- new employees
- liberal workers' compensation laws
- absence of safety laws
- absence of safe work practices & safety programs

Experience has shown that a lack of well-managed safety programs and safe work practices are major factors causing the wastewater treatment industry to belong to an industrial group ranking near the top of the NSC's worst industries with regard to worker safety. Because of these findings, it is the last item listed, the absence of safe work practices and safety programs, that this text is designed to address.

One might ask, if the wastewater industry has such a high incidence of on-the-job injury, why will well-managed safety programs and safe work practices make a difference? To begin with, workers involved with wastewater work have a high incidence of injury because of the *diversity* that is required of them when they perform their assigned duties. The average wastewater worker must be a Jack or Jill of all trades. For example, operating the plant is one thing. That is, taking samples, operating, monitoring, and determining settings for chemical feed systems and high-pressure pumps, performing laboratory tests, and then recording the results in the plant daily operating log are routine functions performed by most wastewater operators. It is the non-routine functions that cause the problems. For example, the typical wastewater operator must not only perform the functions stated above, he/she must also make emergency repairs to systems (e.g., welding a broken machine part to keep the equipment online), perform material handling operations, make chemical additions to process flow, respond to hazardous materials emergencies, perform site landscaping duties, and carry out several other functions that are not usually part of the operator's job classification but are necessary to maintain satisfactory plant operation and site appearance. Remember, the plant operator's job is to *keep the plant running* — keeping the plant running at 3:00 A.M. may require the operator to perform mechanical tasks that he/she is not trained to do.

Since the wastewater operator is expected to be a diverse, multitalented, extremely capable individual who can do whatever is required to maintain smooth plant operation, several safety considerations come into play during a wastewater operator's normal plant shift. Thus, there is a need for a wide variety of safety programs and safe work practices to cover a wide variety of diverse job functions. It logically follows that diverse job functions expose the worker to myriad hazards.

Let us now take a look at the safety organization within a typical wastewater treatment plant. In a wastewater treatment facility it is not unusual for the personnel manager or other designated organizational representative to be thrust into the all-encompassing world of safety. This is especially the case in smaller organizations. As a case in point, consider the example of the small wastewater treatment facility that employs less than 50 full-time workers. Along with a chief operator or plant superintendent, the small wastewater treatment facility generally employs someone in the capacity of personnel manager/payroll clerk/timekeeper and safety person.

Personnel functions, including time-keeping and payroll accounting, are difficult enough in themselves. Add safety to the mixture, and the ingredients do not always easily blend. Indeed, safety is not only a science in its own right, it is also an endeavor that requires full-time attention. The average person who might be thrown into the above predicament may have no or very limited knowledge of safety. As a matter of fact, this same individual may have difficulty in properly explaining the term OSHA (this shortcoming is resolved quickly, however, whenever the organization is cited by OSHA).

The primary lesson the wastewater treatment facility "safety person" must learn to be successful is to be an *advocate for* safe work conditions in his/her facility, not just a *regulator of* safe work conditions. Secondly, when the uninitiated person is thrown into the position as the "safety person," he/she must quickly come to grips with the fact that on-the-job injuries are very real and can be frequent occurrences. On the other hand, it can take a lot longer for the "rookie safety person" to realize that Grimaldi and Simonds (1989) put it right when they stated that "many [injurious] events, almost 9 out of 10 that occur in work places . . ., can be predicted" (p. 3). The point Grimaldi and Simonds make is that knowledge exists not only on how to predict injuries but also on how to prevent their occurrence.

Where do I start? This is a natural question for the new "safety person" to ask. Typically, as previously stated, someone is assigned the additional duty of safety officer for his/her plant as a collateral duty. It is not unusual to find senior plant operators or chief operators who have been assigned this collateral duty. It would be difficult to find a more challenging or more mind-boggling collateral duty assignment than that of "the safety person job." This statement may seem strange to those managers who view safety as a duty that only requires "someone" to keep track of accident statistics, to conduct plant safety meetings, and perhaps place a safety notice or safety poster on the plant bulletin boards. The fact of the matter is, in this age of highly technical safety standards and government regulations, the safety person not only has much more to do than place posters on the bulletin board, he/she also has much more responsibility.

Beyond answering the question: Where do I start?, this text will provide a guide on how to maintain the safety effort after it is on line. Moreover, it will point out the pitfalls and failures that have been experienced and the lessons that have been gained through years of safety experience. This is not to say that any one person has all the answers; no one does! Safety knowledge is something that has to be gained through a commonsense approach blended with years of experience. Experience can only be gained through doing. The "safety person" has to be a *DOER*. A key element that can aid the DOER in doing what is correct is a dogged determination to make his/her workplace the safest possible.

The other major elements of the safety profession can be learned by reading several of the outstanding texts that are available on the subject. The safety person can never stop learning. Simply stated, one cannot learn all there is to know about safety. One can only try to learn the main factors involved with preventing injuries. When you get right down to it, isn't preventing injuries what safety is all about?

So for the personnel manager, wastewater operator, chief operator, maintenance operator or other wastewater person who suddenly finds him/herself assigned to the prestigious but absolutely demanding collateral or full-time duty as plant safety official, this text is designed for you. Additionally, the good news is, this text is designed to explain a technique-procedure, a paradigm or model that actually works; it has been tested.

Because the methodology described in this text has been used successfully in wastewater treatment, it will provide answers to several questions. For example, what types of safety programs are needed at a wastewater treatment facility? What are the health and safety concerns that are unique to the wastewater industry? What are the applicable regulations? Which safe work practices should be used in the wastewater industry? When one determines which plant safety programs to implement, how does one maintain these? And, how does one measure the results?

There is one question this text does not answer; that is, how do you know when you are finished or when the job has been completed? An effective safety effort is never finished; it is never completed. It never stops. It continues to flow like the water in a large stream. Sometimes the stream flow is interrupted by obstacles in the way. Sometimes the flow is contaminated by unsafe acts. Sometimes the flow comes to a complete stop because of a dam placed in its way by some unenlightened plant official. This is where the designated safety official must step in to freeup the flow. To free the flow, the safety official must be armed with facts.

This text is designed to provide the designated safety person not only with the facts but also lessons gained through years of making good judgments and some not-so-good judgments. Moreover, although it is true that there are several outstanding safety texts available, it is also true that many of these texts are targeted for use by officials who have some background in safety. This text, on the other hand, is user friendly — it is written to be used by that vast majority of uninitiated individuals who might find themselves assigned the dubious task of plant safety official. In other words, this text is designed for those individuals who might ask the question: Where do I start?

CHAPTER 2

Safety Starts at the Top

Section 5 (a) of Public Law 91-596 of December 29, 1970, Occupational Safety and Health Act (OSH Act) requires

Each employer—

> (1) Shall furnish to each of his employees employment and a place of employment which are free from recognized hazards that are causing or are likely to cause death or serious physical harm to his employees;
>
> (2) Shall comply with occupational safety and health standards promulgated under this Act.

No matter how many regulations, standards, and laws are designed to ensure workers' safety and no matter how experienced and/or motivated the plant's designated safety official is, he/she is powerless without strong support from the highest levels of plant management. Simply put, without a strong commitment from upper management, the safety effort is doomed. On the other hand, when organizational management states that it is the company objective to place "safety first," even before productivity and quality, then the proper atmosphere is present for the safety official to accomplish the intended objective. That is, the safety official will be able to provide a safe place for all employees to work in.

When assigned the duty as the plant's safety official, the first thing to do is to meet with upper management and determine what the safety objective is. The newly appointed safety official should not start on the plant's safety program until he/she is certain of what is expected of him/her. The pertinent matters that need to be addressed during this initial meeting include developing a written *safety policy* and *safety budget* and determining exactly what *authority the safety official has* and *to whom he/she reports*. In

7

addition, the organization's safety rules and safety committee structure must be formulated as soon as possible.

THE ORGANIZATION'S SAFETY POLICY

The plant safety official should propose that an organizational safety policy be written and approved by the general manager or other top plant manager. A well-written organizational safety policy should be the cornerstone of organization's safety program. There are several examples of safety policies used by Fortune 500 companies and others to model your own plant's safety policy after. The key to producing a powerful, tell-it-like-it-is safety policy is to keep it short, to the point, and germane to the overall goal. Many organizational safety policies are well written but are too lengthy, too philosophical. The major point to remember is that the organization's safety policy should be written not only so that it might be understood by every employee, but also so that all employees will actually read it. An example of a short, to the point, and hard-hitting safety policy is provided in Figure 2.1.

NO JOB IS SO CRITICAL AND NO SERVICE

IS SO URGENT -- THAT WE CANNOT

TAKE THE TIME TO PERFORM OUR

WORK SAFELY.

While it is true that the major emphasis is on efficient operations, it is also true that this must be accomplished with a minimum of accidents and losses. I cannot overemphasize the importance that the District places on the health and well-being of each and every employee. The District's commitment to occupational health and safety is absolute. HRSD's safety goal is to integrate hazard control into all operations, including compliance with applicable standards. I encourage active leadership, direct participation, and enthusiastic support of the entire organization in supporting our safety programs and policies.

_____
                    General Manager

FIGURE 2.1. Sample of an organizational safety policy statement used by Hampton Roads Sanitation District (HRSD), Virginia Beach, Virginia. HRSD's safety policy is similar to several other safety policies that are currently being used in wastewater and other industries. The powerful effect of HRSD's safety policy is its brevity. The major point is made in as short of verse as is necessary; thus, this is the type of safety policy that will be read by the employee. More importantly, HRSD's safety policy sends the desired message.

THE SAFETY BUDGET

The safety budget is critical. No one ever said that safety is inexpensive; it is not. To the contrary, it is not unusual for safety divisions to expend six-figure budgets per year on safety and health programs and equipment for wastewater organizations with large work forces. On the other hand, several smaller wastewater treatment facilities, where money for safety is either hard to find or is nonexistent, the total safety budget might be limited to a few hundred dollars per year (this can also be the case in larger wastewater treatment facilities — this is understandable in the present economic climate where money is tight).

As an example of budgeting woes, consider wastewater treatment facilities that are located in rural and sparsely populated counties. It is not unusual for these small Publicly Owned Treatment Works (POTWs) to have a total budget of not more than $200 per year for the POTW's safety programs. Funding the POTW's operation sometimes takes a back seat to other more urgent county requirements. More importantly, when you consider that the cost of an environmental air monitor can cost $2,500 or more, it is not difficult to understand why many POTWs do not possess even one of these critical air-monitoring devices. This is the case even though air monitoring is required to meet various regulations and for confined space entry operations.

When it comes to budgeting, management is concerned with the bottom-line. It is often difficult for management to discern the value of safety in terms of the cost-benefit relationship. Therefore, it is one of the safety official's functions to enlighten management about the significance of a sound safety program and how it relates to the organization's bottom-line. This is not an easy undertaking. Additionally, making an argument for funding that does not exist can be extremely frustrating.

Sometimes the plant safety official is able to convince those who control finances to budget more money for safety if he/she is able to present a strong case or argument. For example, the safety official must make the point that it is less expensive to incorporate safety into the organization than it is to pay for the loss of life, serious injury, hazardous materials incidents that affect the public, medical expenses, destruction of property, employees' lost time, workers' compensation expenses, possible violation of the plant's operating permit, and citations issued by OSHA or other regulators.

The plant safety official who is attempting to increase funding for safety must understand from the very beginning that when organizational money is spent, upper management wants results. It wants to see what its money has bought. For the designated safety official this is a critical area. Some would call it "blowing your own horn." In reality, it should be called communicating success to the extreme. For example, when OSHA inspects one of the organization's facilities and can find little wrong—nothing that can be cited—then this information must be passed on to upper management. Upper management must get the message that its commitment to spending money on safety has paid off; it has saved money; it has prevented fines and embarrassment. More importantly, a strong commitment to an effective safety program can and will prevent fatalities and injuries. When talking about the organization's bottom-line, the safety official must convince upper management that the organization's real bottom-line is the health and well-being of its employees. If upper management's bottom-line is putting financial gain before protecting employees, then it does not need a safety official and the appropriate safety funding; instead, it needs very deep pockets to pay for very expert legal counsel.

SAFETY OFFICIAL'S AUTHORITY

The safety official's authority is important. The degree and extent of this authority is also important. For example, the safety official must have the authority to conduct in-house audits. Such audits are designed to reveal unsafe conditions and/or practices. More importantly, these audits must be followed up. In other words, the safety official must have the authority and upper management's backing to ensure that supervisors correct deficiencies that are found during the audit.

The safety official must have the authority to shut down work in progress, on the spot, whenever unsafe work practices or unsafe conditions are discovered. Although this type of authority is important, it should also be stressed that this is latent authority; authority that is reserved and only to be used with great discretion. Remember, the safety official must take on the facade of being "Good Neighbor Sam," not that of a Gestapo agent.

The safety official must also have the authority to convene plant-wide mandatory training sessions. Training is at the heart of safety. Employees cannot be expected to abide by safe work practices unless they have been properly trained on what is required of them.

Safety training is more effective when the training is provided by those who are "expert" in the subject matter and who take a personal approach. The organization's safety official must be viewed by supervisors and employees as part of the organization. This can be accomplished if he/she takes an active role in learning the plant operation.

In large organizations where there may be several employees working at several different locations, it is important to train supervisors so that they will be able to recognize safety hazards and take the correct remedial actions. Additionally, well-trained supervisors should augment the safety official's effort in providing employee safety training. Training provided by the worker's immediate supervisor, who is competent in the subject matter, is often more effective than training provided by other officials.

The importance of using supervisors in safety training cannot be overstated. The supervisor's importance in this vital area can be seen when one considers that in order to train workers on the proper and safe performance of assigned duties, input from the technical expert (the supervisor) for each job function is critical. Moreover, when supervisors are asked for their input and advice in formulating safe work practices, they generally *buy into* the overall safety program. Thus, these same supervisors often become valuable allies of the safety official and strong supporters of the organization's safety effort.

ACCIDENT INVESTIGATIONS

The plant safety official must have the authority to properly perform other functions. For example, accident investigations are important. Accident investigations can turn up causal factors that point to a disregard for established safety rules and/or safe work practices, or disobedience of direct orders. All these items are going to point the finger of blame on someone. Caution is advised here; the safety official must tread a fine line in this area. His/her intention should be to determine the cause, recommend remedial action, and follow up to ensure that corrective procedures have been put in place. The safety official should never act or perform investigations in a Gestapo-type manner, but must be professional, tactful, efficient, observant, and thoughtful. Also, the safety official should never target individuals for blame. Dragnet's Joe Friday said it best: "Just the facts, ma'am." The organizational safety official should stick to and report only the facts.

SAFETY RULES

One of the first items on the newly assigned safety official's agenda should be the generation and incorporation of the organization's safety rules. However, before submitting a list of safety rules to higher authority for approval, the safety official should think through what he/she is proposing. Rules are everywhere. All through our lives we have functioned according to some set of rules. Workers generally do not like rules. This is especially the case when the rules are unclear or arbitrary. In putting the plant's safety rules together it is wise for the plant safety official to abide by the old acronym and saying: *KISS—Keep It Simple Stupid.*

Safety rules should be straightforward, easily understood, and limited to as few as possible. Concocting volumes of complicated safety rules will result in much wasted effort and another "dust collector" for the shelf. Employees will not follow or abide by rules that occupy voluminous manuals. Additionally, supervisors will have difficulty in enforcing too many rules.

The best safety rules are those that can be read and understood in short order. For example, the organizational safety rules shown in Figure 2.2 are utilized by Hampton Roads Sanitation District. These rules work; they are followed by both supervisors and workers. They are effective for two major reasons. First, they are limited in number and easily understood. Secondly, they are printed on cards designed for billfold occupancy. Thus, they are a ready reference for the employee. Additionally, a copy of the rules is placed at strategic locations within each work center. The work center safety rule posters are printed in larger, bolder print on 10 × 12 inch rigid poster board.

The safety rules shown in Figure 2.2 would not be effective unless the organization provided enforcement. It is best to provide enforcement in the form of both punishment and praise. Depending upon the severity of the infraction, punishment should be provided to those employees who disregard or disobey safety rules. Moreover, when a good safety effort is observed, praise should be provided in the form of letters of commendation and appropriate notations on the worker's performance record.

SAFETY COMMITTEE/COUNCIL

When working with upper management to formulate the organization's safety program, it is important to set up a safety committee or council. Such a committee can provide valuable assistance to the plant safety official. The safety committee should be composed of a cross section of the organization's workforce. Additionally, it should consist of a combination of senior managers as well as employees at mid-grade supervisory levels. If the organization is unionized, a designated union representative must also be assigned.

When formulating a committee, a written mission statement should be prepared. The mission statement should clearly state the goal of the committee and its authority. Safety committee procedures, frequency of meetings, agenda, records to be kept, and a clear line of communication between the committee and top management should also be established.

As stated earlier, a proper functioning organizational safety committee can provide valuable assistance to the organization's safety official. For example, when a new OSHA regulation is issued, it is wise to have manage-

SAFETY RULES

Employees Should Familiarize Themselves
with and Observe All Safety Rules

1. Wear hard hats at all times in designated areas.

2. Wear safety glasses when chipping, hammering, pounding,
    cutting grass, grinding, and using power tools.

3. Wear appropriate safety goggles and face shield when
    handling chemicals.

4. Wear safety goggles and dust masks when entering
    incinerator or when handling ash.

5. Wear safety shoes.

6. Wear safety harness and lifeline when working in or
    around an enclosed manhole, pipe, tank or wetwell where
    there is no other adequate protection against falling
    or retrieval.

7. Wear life jacket when working in or around an open
    manhole, pipe, tank or wetwell containing sewage or
    sludge where there is no other adequate protection
    against drowning.

8. Test for a safe atmosphere with gas detection equipment
    before entering an enclosed manhole, pipe, tank,
    wetwell, or any area subject to explosive, oxygen
    deficient or toxic atmosphere.

9. Wear self-contained breathing apparatus when entering
    an enclosed manhole, pipe, tank, wetwell or any area
    subject to an oxygen deficient or toxic atmosphere not
    proven safe by testing.

10. Ventilate with portable ventilation blower before
    entering an enclosed manhole, pipe, tank or wetwell.

11. Place barricades around all open access hatches,
    manholes, pipe trenches and excavations left
    unattended.

12. Wear appropriate goggles, face mask, and gloves when
    burning or welding.

13. Wear appropriate eye and hand protection when opening a
    hot incinerator access or air door.

14. Wear hearing protection where designated.

FIGURE 2.2. Sample safety rules.

ment and key plant personnel "buy in" on the new requirement before trying to implement it. Taking a new regulation and force-feeding it into an organization is not recommended.

A well-organized and well-intentioned safety committee can provide additional help to the safety official in the form of constructive criticism. For example, whenever decisions have to be made concerning safe work practices, results of safety audits, review of accident investigations, or other safety issues, it is wise to listen to differing points of view. Sometimes safety officials tend to run with the ball before they know in which direction to travel. Establishing consensus after having thoroughly discussed/argued the issue pays large dividends later when attempting to install new ideas into the organization.

WORKER INPUT

Safety officials sometimes overlook a valuable resource that is always present in any organization: *workers*. Some would argue that workers not only make up the organization but that workers are the organization. Safety officials are hired primarily to protect workers from injury. Safety officials sometimes forget their mission; that is, they forget that their primary task is to ensure that workers have a safe place to work.

Workers also have a role in their own safety. When workers hear this statement, they might be surprised. At the beginning of this chapter, part of the OSH Act was stated; the part dealing with the employer's responsibilities under the Act. It may surprise many people to learn (it almost always surprises workers) that the OSH Act mandates the following:

> Section 5 (b) mandates that each *employee* shall comply with occupational safety and health standards and all rules, regulations, and orders issued pursuant to this Act which are applicable to his own actions and conduct.

Several organizational safety programs, policies, manuals, or directives specifically define who is responsible for safety. When the reader reviews such programs, policies, manuals, or directives it usually is clear who has been made responsible for safety. But, on the other hand, when personnel are "designated" as being responsible for safety, the reader might wonder about the rest of the organization's personnel; that is, those personnel who are not "designated" responsible for safety? Shouldn't everyone share this responsibility?

Providing a safe place to work in can be better accomplished when all organizational personnel have input, especially the workers—the rank and file. Input can be in the form of discussion between the worker and the supervisor. It can also be in the form of discussion between the worker and the work center's safety committee member. On occasion, workers provide input directly to the safety official.

The form of safety input is not important. The important thing is to *get the input*. An organizational safety program should encourage worker empowerment. Worker empowerment provides for worker input. For example, input received from workers during an accident investigation where a serious injury occurred due to a workplace hazard is important to the investigating official for formulating his/her final report and recommending remedial action. On the other hand, if this same worker input had been provided to the work center supervisor or safety official prior to the mishap, then proper remedial action might have been effected, thus preventing the mishap from having occurred in the first place.

The question is how does the worker provide workplace hazard information to the safety official? There are several ways to accomplish this. Input to the organizational safety official can be made through the work center's safety committee, for example. Another technique or device is known as an *Unsafe Work Condition Report*. Figure 2.3 shows an Unsafe Work Condition Report that is used at Hampton Roads Sanitation District. This report has been invaluable in providing a means of communicating on-the-job hazards and/or unsafe conditions that workers have observed. Generally, the Unsafe Work Condition Report is routed through the work center chain of command. When this type of safety hazard information is brought to the attention of the supervisor, corrective action usually follows in quick order. If the hazard is an item that must be budgeted for (i.e., it is not an inexpensive quick-fix item), then the supervisor should inform the worker of this matter.

ACCIDENT REPORTING

The organizational safety official should ensure that supervisors prepare and maintain records for reporting worker injuries and accident investigations. In the wastewater industry, where exposure to various microorganisms (some that are harmful to man) in the wastewater stream is a routine occurrence, even the most minor scratch, cut, and abrasion should immediately be reported to the supervisor. Some would say that it is burdensome for the supervisor and the safety official to process paperwork that details minor on-the-job injuries. This might be the case in other industries, but in wastewater treatment plants, it is important to require employees to report all injuries, no matter how minor.

| HAMPTON ROADS SANITATION DISTRICT | UNSAFE WORK CONDITION REPORT |
| | HAMPTON ROADS SANITATION DISTRICT |

| Work Center | Date of Report | Time of Report | AM |
| | | | PM |

Location of Unsafe Work Condition

Nature of Unsafe Work Condition

DESCRIPTION OF UNSAFE WORK CONDITION

1. What is dangerous or unsafe?

2. What tools, machines, materials, etc., are involved?

3. What person(s) is acting in an unsafe manner? Identify

4. What could happen if the unsafe work condition is not corrected?

5. How could an employee(s) injury occur?

6. How could property damage occur?

FIGURE 2.3. Unsafe work condition report form (*Source:* Hampton Roads Sanitation District; used by permission).

7. How long has the unsafe work condition existed?		
8. What should be done to correct the unsafe work condition?		
9. Employee's Personal Comments		
Signed	Title	Date

Supervisor's Personal Comments		
Signed	Title	Date

Employee-Supervisor Safety Committee's Comments		
Signed	Title	Date

Action Taken	
	Date

FIGURE 2.3 (continued). Unsafe work condition report form (*Source:* Hampton Roads Sanitation District; used by permission).

17

Figure 2.4 shows an example of an Employer's First Report of Accident. This form should be filled out by the plant manager within 24 hours following the accident. The plant manager must keep in mind that the report is a fact-finding activity, the main objective being to determine what happened rather than who is at fault. Explanations should be frank and to the point. Avoid using descriptive words such as "unavoidable" and "carelessness" in the explanations.

Figure 2.5 shows the Safety Division Followup Accident Investigation Report used at HRSD. The purpose of this form is to document a formal investigation required by the insurance vendor when lost time, medical care, or worker misbehavior is involved.

The Employer's First Report of Accident and the Safety Division's Followup Accident Investigation forms are important, as they help ensure that proper documentation has been noted, that Worker's Compensation claims are promptly processed, and that proper entries have been made in the OSHA-200 Log. Further, the Employer's or Supervisor's First Report of Accident is critical in providing the work center supervisor, human resource manager, and safety official the information necessary to correctly document mishaps. Remember, OSHA not only requires the entry of job-related mishaps and illnesses in the organization's OSHA-200 Log, it also requires posting of the final sheet of the log with cumulative totals for the preceding year from February 1 thru 1 March. These postings must be made clearly visible and accessible to all employees. The OSHA-200 Log will be covered in greater detail later in this text.

One additional word of caution should be mentioned here with regard to how the supervisor or assigned safety official fills out the accident and investigation forms. These forms, including the person filling out these forms, might end up in a court of law. It is not unusual, especially in this age of "let me sue you before you sue me," that a job-related mishap, injury or fatality might entail future litigation. This is an important factor that all investigators should keep in mind.

SAFETY AUDITS

Safety audits or inspections can be a valuable tool in detecting work site hazards that may lead to worker injury. The obvious purpose of safety audits is to identify and correct workplace hazards. Not surprisingly, a newly assigned safety official is sometimes apprehensive about conducting safety audits of his/her own organization's facilities. This apprehension stems from fear of antagonizing the site supervisor. Moreover, unless the "rookie" safety official has previous safety inspection experience, he/she

EMPLOYER'S FIRST REPORT OF ACCIDENT
(Every question must be answered)

Employee's Name ....

I.C. File No. ....

Employer

1. Name of Employer .... Phone No. ....
2. Address: No. and St. .... City .... State .... Zip ....
3. Location, if different from mail address ....
   Parent company or policy name ....
4. Insured by: Name of Company .... Policy No. ....
5. Nature of business (or article manufactured) .... Federal Tax I.D. No. ....

6. (a) Location of Plant or place where accident occurred ....
   .... (City or County)
   .... State if employer's premises ....
   (b) If injured in a mine, did accident occur on surface, underground, shaft, drift or mill ...

Time and Place

7. (a) Date of Injury .... 19 .... Day of week ...... Hour of day ...... A.M. ...... P.M.
   (b) Was injured paid in full for day of injury? ....
8. (a) Date incapacity began .... 19....... A.M. ...... P.M.
9. Was injured paid in full for day incapacity began? ....
10. When did you or foreman first know of accident? ....
11. Name of foreman ....

Injured Person

12. Name of injured ....
    (First Name)   (Middle Name)   (Last Name)   (Social Security No.)
13. Address: No. and St. .... City .... State .... Zip ....
14. Check (✓) Married ...... Single ...... Widowed ...... Divorced ...... Male ...... Female ...... No. of Dependent Children ....
15. Date of birth .... Did you have on file employment certificate or permit? .... Employee's Phone No. ....

FIGURE 2.4.

19

16. (a) Occupation when injured .... (b) Was this his or her regular occupation? ....
    In what department regularly employed? ....
17. (a) Date of Hire .... ; How long in present job? .... (b) Place or time worker ....
    (c) Wages per Hour $ ....
18. (a) No. hours worked per day .... (b) Wages per day $.......... (c) No. days worked per week ....
    (d) Average weekly earnings .... (e) Work starts on .... and ends on .... (f) Time shift started
    A.M. .... P.M. (g) if board, lodging, food or other advantages furnished in addition to wages, give estimated value per day, week or
    month ....

Cause of Injury

19. Machine, tool or thing causing injury .... 20. Kind of power, (hand, foot, electrical, steam,
    etc.) .... 21. Part of machine on which accident occurred
22. (a) Was safety appliance or regulation provided? .... (b) Was it in use at time? ....
23. Was accident caused by injured's failure to use or observe safety appliance or regulation? ....
24. Describe fully how accident occurred, and state what employee was doing when injured
    ....
25. Name and address of witness ....

Nature of Injury

26. Nature of injury (describe exact location of amputation or fracture, right or left) ....
    ....
27. Probable length of disability .... 28. Has injured returned to work? ....
    If so, date and hour .... At what wage $ ....
29. At what occupation? ....
30. (a) Name and address of physician ....
    (b) Name and address of hospital ....

Fatal Cases

31. Has injured died? .... If so, give date of death ....

Date of this report .... Firm Name ....
Signed by .... Official Title ....

FIGURE 2.4 (continued).

HAMPTON ROADS SANITATION DISTRICT
SAFETY DIVISION'S
ACCIDENT INVESTIGATION REPORT

I. GENERAL INFORMATION

EMPLOYEE NAME _____

SOCIAL SECURITY NUMBER _____

SEX _____ DATE OF BIRTH _____ JOB TITLE _____

FACILITY _____ DEPARTMENT _____

DATE AND TIME OF ACCIDENT _____

EXACT LOCATION OF ACCIDENT _____

II. NATURE OF INJURY/ILLNESS _____

NATURE OF INJURY AND PART OF BODY _____

TREATMENT: FIRST AID _____ MEDICAL _____

PHYSICIAN: NAME _____ ADDRESS _____ PHONE _____

HOSPITAL NAME _____ ADDRESS _____ PHONE _____

FIGURE 2.5. Safety division's followup accident investigation report.

21

SEVERITY OF INJURY: FATALITY _____
                    LOST DAYS (DAYS AWAY FROM WORK) _____
                    LOST WORKDAYS (DAYS OF RESTRICTED ACTIVITY) _____
                    OTHER (SPECIFY) _____

DAMAGE TO PROPERTY OR EQUIPMENT _____

III. DESCRIPTION OF INCIDENT

WHAT HAPPENED AND HOW DID IT HAPPEN: _____
_____
_____

EMPLOYEE INVOLVED (COMMENTS): _____
_____
_____

WITNESSES (NAMES & COMMENTS): _____
_____

SUPERVISION AT TIME OF ACCIDENT: _____
_____

IV. ANALYSIS

WHAT CAUSED THE INCIDENT? WHY DID IT HAPPEN? _____
_____
_____

CONTRIBUTING FACTORS: _____
_____

V. PREVENTATIVE/CORRECTIVE ACTION

FIGURE 2.5 (continued). Safety division's followup accident investigation report.

IMMEDIATE _____

_____

_____ COMPLETION DATE(S) _____

WHO IS RESPONSIBLE _____

LONG TERM _____

_____ COMPLETION DATE(S) _____

WHO IS RESPONSIBLE _____

VI. SUMMARY

CAN EMPLOYEE RETURN TO WORK WITH COMPETENT PHYSICIANS APPROVAL? _____

IF YES TO WHAT EXTENT (LIGHT DUTY OR REGULAR DUTY) _____

_____

PHYSICIANS COMMENTS: _____

_____

_____

COMPLETED BY: _____ DATE: _____
TITLE: _____ DIVISION: _____

REVIEWED BY: _____ DATE: _____
TITLE: _____ DIVISION: _____

FIGURE 2.5 (continued). Safety division's followup accident investigation report.

23

often feels that he/she lacks knowledge and expertise in conducting safety inspections.

One of the questions that the new safety official might ask is: What do I inspect for? While it is true that each work site is unique, it is also true that safety hazards generally fall into the same categories, no matter the size of your facility. There are exceptions to this rule, however. For example, the hazards that are generic to a nuclear facility are not the same as the hazards that are present in a wastewater treatment plant.

Concerning questions about conducting safety audits in wastewater treatment plants, the inquirer should refer to several excellent publications that discuss the topic in detail. Several of these publications are listed in the reference section of this text. Additionally, the potential auditor should do a number of things prior to conducting the audit. First, the auditor should be familiar with the facility and operation to be audited. Moreover, he/she should gain familiarity with the types of hazards that are normally present at the facility. For example, the auditor should determine whether the facility handles and uses hazardous materials.

Wastewater treatment plants possess most of the industrial hazards that are present in other industrial settings. A short list of some of the typical hazards that are present at wastewater treatment plants is provided as follows:

(1) Machinery
(2) Flammable-combustible materials
(3) Walking-working surfaces
(4) Welding, cutting, and brazing operations
(5) Electrical equipment and appliances
(6) Ladders and scaffolds
(7) Compressed gases
(8) Materials handling and storage
(9) Hand and portable powered tools
(10) Process-generated hazardous and toxic substances (e.g., methane and hydrogen sulfide)

A valuable tool when conducting the safety audit is a Safety Inspection Checklist. If your organization does not have a safety inspection checklist, one should be generated as soon as possible. Once the organization's safety checklist is generated, the safety official should consider this document as a "living" document that will continue to grow as time passes.

Armed with the safety checklist, the safety official is ready to seek out the senior person present and conduct the audit. The supervisor in charge should always be asked to accompany the safety auditor on the inspection. When the senior person at the facility is present during the inspection, the inspection takes on added importance. In addition, when the senior person present accompanies the inspector, he/she is able to discuss any discrepancies that the inspector might find. This is important. When the final inspection report is published for corrective action, it is important that there are no surprises that would embarrass and thus infuriate the senior person in charge. Any disagreement with the inspector's findings should be worked out with the supervisor who is responsible for the work center prior to the formal publishing of the results.

Many discrepancies found during safety audits can be remedied on the spot. It should be remembered that after having discovered a hazard or hazards, it is the prompt and correct remedial action that is the ultimate goal of the safety audit. Figure 2.6 shows a safety audit/inspection checklist that is used at HRSD. This checklist does not include every possible hazard. To do so, the list would have to be a voluminous publication. This is what you are attempting to avoid. Instead, it is wiser to generalize some hazards into categories. For example, trip, slip, and fall hazards cover a wide general area. These hazards can be specifically itemized and characterized later in the comments section of the report.

COMMUNICATION

Throughout this text the importance of training and documentation of the training that has been completed will be stressed. Another area is equally as important as training workers and documenting the training. This is *Communication*. Communication is addressed in OSHA's Hazard Communication (HAZCOM) Standard. The HAZCOM Standard requires employees who might be exposed to hazardous materials to have full knowledge of the hazards. Safety communications goes beyond HAZCOM. Getting safety information out to the supervisor and worker is critical. For example, in order to facilitate scheduling of safety training it is prudent to publish a schedule far in advance of the actual training dates.

In order to give plenty of advance notice of future safety training, it might be wise to develop an organizational publication, *Safety Training/Medical Exam/Inspection Schedule,* that can be published quarterly or annually. This schedule should be a day-by-day, month-by-month account of when safety training is scheduled. Supervisors like to know far in advance what is scheduled because it allows them to plan ahead.

Another area of communication that is important to safety is getting the word out about new safety requirements, regulations, safe work practices or

HAMPTON ROADS SANITATION DISTRICT
SAFETY INSPECTION CHECKLIST

INSPECTION ITEM/S:	SAT	UNSAT	COMMENTS
1. Environmental factors (illumination, noise, vapors, etc)			
2. Hazardous materials storage/labeling			
3. Machinery safety guards			
4. Electrical equipment			
5. Portable tools			
6. PPE (safety glasses, hardhats, safety shoes)			
7. First Aid supplies, safety showers and eye-washes			
8. Fire protection equipment (extinguishers and hoses)			
9. Walkways and roadways			
10. Elevators			
11. Working surfaces (ladders, scaffolds, catwalks, platforms)			
12. Material handling equipment (chains, slings, hoists, dollies)			
13. Railroad tank cars			

FIGURE 2.6. A safety audit/inspection checklist used by HRSD to audit 9 wastewater treatment plants, 76 pumping stations, 2 large maintenance facilities, and a compost site. This checklist is completed by the auditor and one copy is forwarded to the work center supervisor, one copy to the senior departmental representative, and one copy is maintained on file by the safety official.

INSPECTION ITEM/S:	SAT	UNSAT	COMMENTS
14. Warning and signaling devices			
15. Containers			
16. Storage facilities			
17. Structural openings (doors and windows)			
18. Work procedures and practices			
19. Retaining walls around chemical storage areas.			
20. Right–to–Know stations, MSDS, labeling, stenciling			
21. Wind socks in vicinity of Cl2/SO2 storage			
22. Emergency lighting			
23. Nighttime lighting			
24. Respirators and records			
25. CL2 system Integrity (pigtails. etc.)			
26. Spill clean–up equipment (booms, absorbent)			
27. Check vent systems			
28. Drum storage (flammable and combustible liquids)			
29A. CL2 repair kits/CL2 gaskets			
29B. SO2 repair kits – SO2 gaskets			

FIGURE 2.6 (continued). A safety audit/inspection checklist used by HRSD to audit 9 wastewater treatment plants, 76 pumping stations, 2 large maintenance facilities, and a compost site. This checklist is completed by the auditor and one copy is forwarded to the work center supervisor, one to the senior departmental representative, and one copy is maintained on file by the safety official.

27

INSPECTION ITEM/S:	SAT	UNSAT	COMMENTS
30. Safety Training records			
31. Workcenter employee Bulletin Boards — (Check for Safety postings & Equal opps., Worker Comp., & OSHA's Regs.)			
32. CL2 Vent Auto — Shutdown			
33. Unsafe Work Condition Form			
34. 4' Toeboards installed with STD railing			
35. CL2 — SO2 in compliance with sewerage regs. segregation/marking off requirements			
36. Walking surfaces modified for icy conditions			
37. Guardrail chain taut			
38. Guardrail chain midrail			
39. Guardrail chain 200LB req.			
40. CDL driver's check			
41. PQS			
42. Electrical SWBD Circuit Labels			

FIGURE 2.6 (continued). A safety audit/inspection checklist used by HRSD to audit 9 wastewater treatment plants, 76 pumping stations, 2 large maintenance facilities, and a compost site. This checklist is completed by the auditor and one copy is forwarded to the work center supervisor, one to the senior departmental representative, and one copy is maintained on file by the safety official.

INSPECTION ITEM/S:	SAT	UNSAT	COMMENTS
43. Equipment Identification (Stencils)			
44. Electrical grade matting			
45. Out of service signs (Electrical)			
46. Hanging valve chains (Marked)			
47. Valves that are reached by ladder			
48. Valve stems covered			
49. Eyewash bottles within 10 secs.			
50. Eyewash valves chained open			
51. Oil racks — capacity labeled			
52. Fixed ladders			
53. Forklift nameplates			
54. Acetylene gage red-lined @ 15 psi			
55. Cutting hoses married correctly			
56. Terminals on ARC Welders covered			
57. Welding cable — 10' to stingers			

FIGURE 2.6 (continued). A safety audit/inspection checklist used by HRSD to audit 9 wastewater treatment plants, 76 pumping stations, 2 large maintenance facilities, and a compost site. This checklist is completed by the auditor and one copy is forwarded to the work center supervisor, one to the senior departmental representative, and one copy is maintained on file by the safety official.

29

INSPECTION ITEM/S:	SAT	UNSAT	COMMENTS
58. Lawnmowers			
59. Files — handles			
60. Jacks — capacity marked			
61. Pedestal grinder anchored			
62. Carpenter shop machines mag cut-off			
63. GFCI's for extension cords			
64. Gas pump emergency shut-off labeled			
65. Propane guarded & graveled			
66. Portable electrical cables			
67. Port fire ext on cutting carts			
68. Gas tech calibration log			
69. Medical Restriction Knowledge			
70. Critical valves marked			
71. Emergency response decision tree			
72. Chemical suits			
73. Employee knowledge Safety SOP			
74. Alcohol Sensor Calibration			

FIGURE 2.6 (continued). A safety audit/inspection checklist used by HRSD to audit 9 wastewater treatment plants, 76 pumping stations, 2 large maintenance facilities, and a compost site. This checklist is completed by the auditor and one copy is forwarded to the work center supervisor, one to the senior departmental representative, and one copy is maintained on file by the safety official.

SAFETY NOTICE #72

The Federal Register published 6/29/95 has strengthened the requirements of CFR 1910. (General Industry) 1001 & CFR 1926. (Construction)1101 Asbestos Standard.

The new requirements mandate the following:

    *lowers the permissible exposure limit

    *requires new worker protection scheme

    *mandates <u>enhanced</u> employee training.

The new requirements have been accepted by Virginia and must be complied with by 10/1/95.

What does all this mean?

1.   All licensed Asbestos workers, supervisors, and project monitors must be trained on new requirements prior to 10/1/95. This training can be conducted in-house.
The Safety Division is currently assembling the new information into a training package that will be presented to all licensed asbestos workers during the following schedule:

    NSMC--0715 September 26, 1995

    SSMC--0715 September 27, 1995

Training sessions will be 1-hour in duration.

2.   HRSD's Asbestos Awareness Training Program for all employees must be upgraded by 10/1/95; this is being accomplished at present.

FRANK SPELLMAN, CSP

FIGURE 2.7. Sample safety notice.

lessons-learned. A *Safety Notice* or *Safety Bulletin* format might be useful (see Figure 2.7). This format is straight to the point and allows for quick and easy dissemination of vital information.

The bottom-line on safety communication is that both supervisors and workers need to be informed. While it is true that too much information can defeat the intended purpose, it is also true that too little information may lead to accidents.

CHAPTER 3

Safety Programs

There are two types of safety programs: organizational safety programs and individual safety programs. An organizational safety program is the plant's *entire* safety program, which consists of policy, organization, and the safety and health plan. Individual safety programs are specific safety programs (e.g., hazard communication, confined space entry, lockout/tagout, respiratory protection, and other safety programs), which are part of the organizational (plant) safety program.

Individual safety programs are designed to achieve specific objectives: To state a plan of action for the enlistment and maintenance of support from all members of each particular area pertaining to safety on the job. In order to be effective, the plant's individual safety programs must be understandable and their requirements must be clearly defined. Moreover, they must be *enforceable.* Without enforcement, safety cannot and will not become part of the organization's culture.

A newly assigned plant safety official might ask: Which individual safety programs should be implemented in my organization? OSHA is specific with regard to safety programs. Moreover, OSHA details specific requirements for several different types of required safety programs. The newly assigned safety official is literally confronted with volumes of material on the subject. For example, the newly assigned safety official can consult sources that include state and national safety councils, insurance companies, police and fire departments, hospitals, and local state health departments.

When wastewater treatment plant safety officials take the first step to determine which OSHA-required safety programs are needed for their particular facility, they must keep in mind that organizational safety programs should be designed to fit the organization and its function.

Wastewater treatment facilities that have four to six employees obviously have different needs than the wastewater treatment plant employing thirty to fifty or more employees.

Some of the OSHA-required safety programs apply to all facilities; some do not. This chapter stresses safety programs that are required for a wastewater facility with more than ten workers. However, during the discussion of each required individual safety program, the safety programs that are required at all wastewater treatment facilities, *no matter their size,* will be pointed out.

In the safety program design phase you must keep in mind that writing a safety program to comply with a particular mandate is only part of the requirement. Employees must be trained on the organization's safety programs. As a matter of fact, more than 100 specific training programs are mandated by OSHA, EPA, and DOT regulations. The key thing to remember is that these training programs are not optional; if your employees perform tasks covered by regulations, you must provide the specified safety training. Before discussing what needs to be accomplished in the training of your workers to meet compliance, it is best to first discuss the safety programs that might be needed for your particular wastewater facility.

SAFETY PROGRAMS FOR WASTEWATER TREATMENT PLANTS: GUIDE TO COMPLIANCE

The first step in designing specific safety programs for your facility is to list the criteria that should be used to test the design effectiveness of each program. This can best be accomplished by implementing safety programs that meet the following criteria:

- Make sure the program is comprehensive.
- Constantly measure the degree of program implementation.
- Ensure that competent persons are trained and assigned as required.
- Ensure that each program is enforceable.

HAZARD COMMUNICATION: THE RIGHT TO KNOW LAW (29 CFR 1910.1200)

If your workers come into contact with hazardous chemicals in your workplace, OSHA mandates that they have a "right to know" what chemicals they are working with or around. This employee "right to know" requirement is formally known as OSHA's *Hazard Communication Standard*. This standard requires that the manufacturers, importers, and distributors of the hazardous chemicals transmit important information to employees in the form of container labels and Material Safety Data Sheets (MSDSs). The information contained on the label or MSDS must clearly state the possible physical or health hazards of each chemical.

Under the Hazard Communication Standard, the *employer* is required to ensure that each chemical is properly labeled. In addition, the employer is required to label hazards that might be produced by a *process*. For example, in the wastewater industry it is not unusual for deadly methane gas to be generated in the wastestream. Another process hazard that is common in the wastewater industry is the generation of hydrogen sulfide gas during degradation of organic substances in the wastestream. OSHA's Hazard Communication Standard requires the employer to label this hydrogen sulfide hazard so that workers are warned of the hazard and that safety precautions are to be followed (see Figure 3.1).

Labels must be designed to be clearly understood by all workers. In addition, employers are required to provide both training and written materials to make workers aware of what they are working with and what hazards they might be exposed to. Employers are required to make Material Safety Data Sheets (MSDSs) available to all employees (see Figure 3.2). The facility Hazard Communication Program must be in writing and, along with MSDSs, made available to all workers 24 hours each day/shift.

Figure 3.2 shows that the MSDS provides seven or more sections on each hazardous material (HAZMAT). Section 1 identifies the HAZMAT and the

CAUTION

HYDROGEN SULFIDE

GAS MAY BE

PRESENT

FIGURE 3.1. Standard warning label.

MATERIAL SAFETY DATA SHEET

Trade Name As Used On Label _____

Manufacturer _____

MSDS Nnumber _____

Address _____

CAS Number _____

Date Prepared _____

Phone Number (For Information) _____

Prepared By _____

Emergency Phone Number _____

Note: Blank spaces are not permitted. If any item is not applicable, or no information is available, the space must be marked to indicate that.

SECTION 1 – MATERIAL IDENTIFICATION AND INFORMATION

COMPONENTS – Chemical Name & Common Names (Hazardous Components 1% or greater;Carcinogens 0.1% or greater)	OSHA PEL	ACGIH TLV	OTHER LIMITS RECOMMENED

FIGURE 3.2. Blank Material Safety Data Sheet (MSDS); front and reverse sides.

Non-Hazardous Ingredients

TOTAL

| SECTION 2—PHYSICAL / CHEMICAL CHARACTERISTICS |

Boiling Point

Specific Gravity (H2O = 1)

Vapor Pressure (mm Hg and Temperature)

Melting Point

Vapor Density (Air = 1)

Evaporation Rate (= 1)

Solubility in Water

Water Reactive

Appearance and Odor

| SECTION 3—FIRE AND EXPLOSION HAZARD DATA |

Flash Point and Method Used	Auto-Ignition Temperature	Flammability Limits in Air % by Volume	LEL	UEL

Extinguisher Media

Special Fire Fighting Procedures

Unusual Fire and Explosion Hazards

FIGURE 3.2 (continued). Blank Material Safety Data Sheet (MSDS); front and reverse sides.

37

SECTION 4–REACTIVITY HAZARD DATA

STABILITY | Conditions To Avoid

Incompatability (Materials to Avoid)

Hazardous Decomposition Products

HAZARDOUS POLYMERIZATION | Conditions To Avoid

SECTION 5–HEALTH HAZARD DATA

| PRIMARY ROUTES OF ENTRY | Inhalation | Ingestion | CARCINOGEN | NTP | OSHA |
| | Skin Absorption | Not Hazardous | LISTED IN | IARC Monograph | Not Listed |

HEALTH HAZARDS | Acute
| Chronic

Signs and Symptoms of Exposure

Medical Conditions Generally Aggravated by Exposure

EMERGENCY FIRST AID PROCEDURES–Seek medical assistance for further treatment, observation and support if necessary.

Eye Contact

Skin Contact

Inhalation

Ingestion

FIGURE 3.2 (continued). Blank Material Safety Data Sheet (MSDS); front and reverse sides.

SECTION 6—CONTROL AND PROTECTIVE MEASURES

Respiratory Protection
(Specify Type)

Protective Gloves | Eye Protection

| VENTILATION | Local Exhaust | Mechanical (general) | Special |
| TO BE USED | Other (specify) | | |

Other Protective
Clothing and Equipment

Hygienic Work
Practices

SECTION 7—PRECAUTIONS FOR SAFE HANDLING AND USE / LEAK PROCEDURES

Steps to be Taken if Material
Is Spilled Or Released

Waste Disposal
Methods

Precautions to be Taken
in Handling and Storage

Other Precautions and / or Special Hazards

| NFPA Rating* | Health | Flammability | Reactivity | Special |
| HIMIS Rating* | Health | Flammability | Reactivity | Personal Protection |

FIGURE 3.2 (continued). Blank Material Safety Data Sheet (MSDS); front and reverse sides.

39

chemical components it contains. Section 2 provides important physical/ chemical characteristics. The hazardous materials emergency responder pays particularly close attention to this section. As a case in point, consider that if your plant site has a chlorine spill and emergency response is required to mitigate the spill. The first item the emergency responder should seek out is the MSDS for chlorine. From the MSDS for chlorine, the emergency responder can quickly determine that chlorine has a vapor density of 2.5. This means that chlorine gas is 2.5 times heavier than air and will seek out any low spots near the spill. Additionally, on the MSDS for chlorine the emergency responder will find that as a gas chlorine is a respiratory hazard that can kill. This could be life-saving information for the responder who might have to rescue someone from an underground vault or storage space.

Section 3 of the MSDS provides important information on the HAZMAT's fire and explosion potential. Obviously, whenever the local fire department is called to the scene of a HAZMAT emergency, it is very interested in the fire and explosive potential of the material spilled. This is especially the case whenever a HAZMAT incident includes fire.

Section 4 details information concerning the reactivity of the HAZMAT. *Reactivity* refers to the stability of the material. Obviously, a HAZMAT such as nitroglycerin is extremely unstable and must be handled with great care. On the other hand, a substance like acetone is fairly stable or nonreactive.

Section 5 covers information related to the health hazards associated with the HAZMAT. For example, important information relative to the primary routes of entry into the body by inhalation, ingestion, and/or skin absorption is covered. Additionally, the MSDS points out whether or not the HAZMAT may cause acute or chronic health problems. Additionally, emergency first aid procedures are listed. This is critical information, especially if a worker is exposed. If the worker's eyes, skin, respiratory system, or internal digestive tract is contaminated with the HAZMAT, the emergency responder needs to know immediately what emergency medical response actions to take.

Personal Protective Equipment (PPE) is listed in Section 6 of the MSDS. It is important for workers to know how to protect themselves from the lethal effects of the HAZMAT. Employees who are working with or around a particular hazardous material need to pay close attention to this section, which details the respiratory protection, eye protection, hand protection, and protective clothing and equipment that might be required to work safely with the HAZMAT. In addition, Section 6 addresses the kind of ventilation that might be required to make the HAZMAT environment safe to work in. Another part of Section 6 that has direct implications for the wastewater industry is the hygienic work practices. Wastewater workers are constantly exposed to the biological hazards associated with handling untreated wastewater. This kind of hazardous exposure potential can be greatly magnified when HAZMATs are thrown into the equation.

The last section, Section 7, points out the correct steps to be taken during a HAZMAT leak/spill. HAZMAT workers need to remember that whenever a HAZMAT is spilled or released to the environment, they are no longer working with a HAZMAT; instead, they are now working with a hazardous waste. Section 7 addresses the proper hazardous waste disposal methods; this is important. Moreover, if there are any special precautions to be taken or hazards to be aware of, Section 7 addresses these items as well.

Before leaving this important discussion of OSHA's Hazard Communication Standard and the requirements for compliance, it is important to point out that no matter what size wastewater treatment facility you work in, you must comply with this standard. This information may surprise a large number of chief operators who work at local POTWs. At a small wastewater facility, the thinking usually evolves around the facility's size. That is, most small facility operators usually think that because they are only a one- or two-person operation they do not use hazardous materials in their processes. In order to check on whether or not your facility must comply with the hazard communication standard, it is necessary to conduct a survey of your facility. During this plant survey a determination should be made as to whether or not hazardous chemicals are used or produced. Remember, a process that does not use hazardous materials but, instead, produces hazardous by-products (i.e., methane or hydrogen sulfide) must be listed (as required in OSHA's hazard communication standard) in the facility's written hazard communication program. At the very minimum, warning signs must be posted alerting workers and visitors to the potential hazard. Remember, the essence of OSHA's Hazard Communication Standard is that the *employee has a right to know.* This right to know also includes *anyone* who might be exposed to the hazard.

CONTROL OF HAZARDOUS ENERGY: LOCKOUT/TAGOUT (29 CFR 1910.147)

Lockout/tagout refers to the control of energy, including electrical, chemical, hydraulic, pneumatic, thermal, and potential energy (e.g., energy that is stored in a compressed spring). Lockout/tagout *procedures* are designed to prevent accidents and injuries caused by the accidental release of energy. All wastewater treatment facilities are required to have a written lockout/tagout procedure. This procedure can prevent needless deaths and serious injuries to workers. The plant's lockout/tagout procedure *must be* the only acceptable method used to deenergize equipment and machinery and control the release of potentially hazardous energy.

In this standard, OSHA mandates training, safety audits, and recordkeeping to ensure that workers will not be unintentionally injured by equipment that is unintentionally energized. The *lockout* procedure is to be accomplished by installing a lockout device(s) at the power source so that equipment powered by that source cannot be operated. A lockout device is a lock or chain or block or some other device (see Figure 3.3) that keeps a switch, valve, or lever in the off-position. Locks are to be provided by the employer and are not to be used for personal use such as to lock tool boxes or employee lockers. Each lock provided to the employee must be employee-identifiable (e.g., marked with employee's name or initials) and individually keyed (only the employee performing the lockout can possess the key to his/her locks). A master key to each employee's locks can be maintained by the plant supervisor. A word of caution is advised here, master keys must be strictly controlled and used only in emergency situations to remove an employee's lock.

A *tagout* is accomplished by placing a warning not to restore energy. Tags must clearly state: *Do not operate. Do not energize. Hands off,* or some other such warning (see Figure 3.4).

When lockout is possible, it is the preferred method used. Although tagout has its applications, it should be used sparingly. Remember, OSHA requires that if a machine or system can be locked out, it must be locked out.

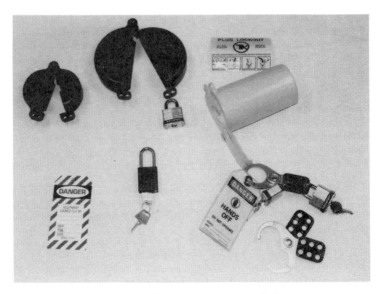

FIGURE 3.3. Sample lockout/tagout devices.

FIGURE 3.4. Sample lockout/tagout tag.

Most workers associate lockout/tagout procedures with deenergizing electrical equipment. It must be remembered, however, that controlling energy sources involves controlling more than just electrical hazards and that the ultimate goal of lockout/tagout procedure is to prevent accidental start-ups, electric shock, and release of stored energy. The following is one example of a lockout/tagout procedure.

Lockout/Tagout Procedure

Before the worker to be protected starts his/her work, the following shall be done:

(1) Notify operations and maintenance supervisors.

(2) Place the main power switch, valve, or operating lever in the *off,* *closed,* or *safe* position.

(3) Check or test to make certain that the proper controls have been identified and deactivated. *NOTE:* After locking or tagging out equipment or machinery and before beginning the work, always try to start the equipment or machine. Obviously, if it starts, it is not properly locked out.

(4) A lock shall be used to secure the disconnection whenever possible. If a lock cannot be used on electrical equipment, for example, an electrician or otherwise qualified person shall remove the fuses to the circuit or disconnect the line leads feeding the equipment.

(5) A hold-off tag shall be attached to the switch, valve or lever. This tag shall bear the name, department, and telephone number of the employee (or division/department) performing the work.

(6) When machine controls or auxiliary equipment are powered by separate power sources, such equipment or controls also shall be locked and tagged to prevent any hazard that may be caused by operating the equipment or exposure to live circuits.

(7) When equipment or processes use hydraulic or pneumatic power, pressure in the lines (or accumulations) shall be relieved. If pressure-relief valves have not been provided, the equipment shall be cycled until the pressure has been dissipated, or the pressure lines shall be opened or disconnected.

(8) When potential or stored energy is a factor as a result of position, spring tension or counterweighting, the equipment shall be placed in the bottom of closed position, or shall be blocked to prevent movement.

(9) When work involves more than one person, additional employees shall attach their locks and tags as they report.

(10) When outside contractors are involved, the equipment is locked out and tagged in accordance with this procedure by the plant project supervisor. Only in emergency cases is equipment to be shut down by other than a plant representative.

CONFINED SPACE ENTRY (29 CFR 1910.146)

Confined space entry and the wastewater treatment and collection industry go hand in hand. If confined space entry is conducted at your facility, then you must have a Confined Space Program. Whether the confined space is a contact tank, aeration basin, clarifier, pumping station wetwell, bar screen trunk, pipe, sewer, manhole, storage vessel, tunnel, boiler, reactor, tank, or a large excavation where repairs to a damaged interceptor or collection line are undertaken, proper confined space entry procedures are essential to protect the worker. Several thousand injuries occur each year in confined spaces. OSHA (1993) defines a confined space as any space that

- has limited or restricted means of entry or exit
- is large enough for an employee to enter and perform assigned work
- is not designed for continuous worker occupancy.

Hazards common to confined spaces include:

- atmospheric conditions
- engulfment
- internal configuration
- physical hazards

The primary factor involved with confined spaces is *oxygen deficiency*. Normal air contains about 20.8 percent oxygen by volume. OSHA specifies that the minimum safe level is 19.5%. According to OSHA, the maximum safe level is 25%. (Remember that an atmosphere too rich in oxygen can cause fire and explosion.)

Combustibility is another factor that makes confined space entry hazardous. Gases and vapors are often trapped in confined spaces. These built-up gases or vapors can easily be ignited by friction, cigarettes, and sparks from hot work equipment.

Another factor that can make a confined space extremely dangerous is *toxic air contaminants*. Toxic air contaminants such as methane, hydrogen sulfide, cleaning solvent, and coating vapors are constant threats to those who work in wastewater treatment and collection. Some toxic air contaminants can cause the entrant problems due to irritation of the respiratory system. Others can be much more serious, they can cut off the entrant's oxygen supply or enter his/her lungs and asphyxiate him/her.

Engulfment is a problem, too. *Engulfment* is defined as the surrounding and effective capture of a person by finely divided particulate matter or a liquid. To gain a better understanding of engulfment and its potential hazard, consider the following example.

If you have ever worked on a farm where grains like wheat and corn are grown, you are probably familiar with grain silos. Grain silos can be extremely dangerous for a number of reasons. First, fine grain dust is always an explosion hazard; second, grain silos are engulfment hazards. The engulfment hazard occurs whenever grain is stored inside the silo and personnel enter the compartment below the storage area. If the grain is suddenly released from the upper compartment and falls or *engulfs* a person below, inside the compartment, survival is doubtful. Industry safety records show several recorded incidents of mishaps whereby farmers and helpers have died due to engulfment in silos.

In the wastewater treatment plant or collection system, engulfment is also a very real possibility. For example, a clarifier that is emptied for maintenance but is not properly locked out could trap the workers inside if the wastestream, under pressure, inadvertently were to enter the clarifier basin and rise in level at more than 10 feet per minute. Additionally, collection crews are at risk if they are working inside an interceptor line that is not properly locked/tagged out and blinded or blanked with flanges. The sudden release of the wastestream into the area they are working in could engulf them. Engulfment can also occur in pumping or lift station wetwells. Additionally, in excavation work where workers are deep inside a trench that suddenly collapses, engulfment can trap, crush or suffocate workers.

Physical hazards are other factors that can make a confined space entry hazardous. Physical hazards include machines like rotating blades or agitators. Moving parts in confined spaces must be locked out/tagged out before entry is made.

Heat stress is another physical hazard that is often precipitated by confined space entry. This should not be surprising when one considers the nature of the atmosphere that is contained in a confined space. Thus, it is not unusual for confined space atmospheres to reach well above ambient temperature. The normal heat sink qualities of some confined spaces are further aggravated by the entrant's presence in the confined space with all the personal protective equipment like self-contained breathing apparatus on his/her person.

Noise is yet another physical hazard that most people ignore or do not think about when confined space entry is made. Noise actually reverberates in confined spaces and can cause permanent hearing loss. Hearing protection may be required to protect the confined space entrant.

Another physical hazard in confined spaces is falls, which can be fatal. Falls are common because confined spaces are usually entered via ladders. Ladders are inherently dangerous and this danger is magnified exponentially when used in confined spaces.

In all incidents where improper confined space entry procedures are used in a space with an undetected contaminant present, there is a common thread. Fatalities are almost always the result. Confined spaces can be very unforgiving. Unfortunately, experience has shown that confined space mishaps result in the deaths of more than one individual. This is the case because of what the author calls the *John Wayne Syndrome*. For instance, when a victim is found unconscious in a wetwell, the person who finds such a victim often attempts heroic rescue efforts to free the victim with total disregard for his/her own safety. Thus, as a result of the John Wayne Syndrome, the confined space tragedy becomes a multivictim incident.

Confined space entry relies on additional safety programs to ensure safe entry. For example, as pointed out earlier, a confined space should not be entered unless proper lockout/tagout procedures have been effected to prevent engulfment and accidental energizing of machinery. In addition to lockout/tagout, another important OSHA safety program that must be followed in order to ensure safe confined space entry is an effective *Respiratory Protection Program.*

It cannot be overstated that in order to ensure safe confined space entry, it is necessary to devise a sound Confined Space Entry Program for your facility. A sound Confined Space Entry Program begins with a written program. Such a program should include entry procedures, lockout/tagout procedures, ventilation procedures, air testing procedures, rescue procedures, personal protective equipment, and confined space permit procedures/requirements for confined space that can only be entered by permit. The confined spaces permit requirements should clearly detail not only how to fill out a confined space permit, but should also designate who is authorized to fill out the permit. All confined spaces must be labeled. OSHA is quite specific about confined space labeling (see Figure 3.5). It is not necessary to expend large sums of money on fancy labels. For example, the confined space permit can be painted on a wall, access hatch or cover. Stenciling the label in heavy black letters is another technique that works.

Thus far, this discussion has addressed confined spaces that are to be entered by permit only. It should be noted, however, that OSHA discusses two types of confined spaces: non-permit and permit-required confined spaces.

CONFINED SPACE
ENTRY BY PERMIT
ONLY!

FIGURE 3.5. Confined space warning sign.

Non-permit confined spaces are those that do not contain physical or atmospheric hazards that could cause death or serious physical harm. Generally, non-permit confined spaces are those that are continuously ventilated, have more than one way in or out, and have no history or hazardous atmosphere. Determining whether a confined space is a non-permit or permit-required space requires professional judgment. *When in doubt, consider the space a permit-required confined space.*

With respect to permit-only confined spaces, OSHA is very specific. These confined spaces are those that contain one or more of the following hazards:

- material that could engulf the entrant (e.g., grain in a silo or wastewater stream in an aeration basin)
- a hazardous or potentially hazardous atmosphere
- an internal shape that could trap or suffocate an entrant
- any recognized serious safety or health hazard

The Confined Space Entry Permit (see Figure 3.6) must be filled out by a competent or qualified person only. Moreover, the Confined Space Permit must be posted at the confined space or be in the custody of the qualified/component person, who must remain at the confined space during confined space entry procedures. This individual must be thoroughly trained, and designated/certified in writing (see Figure 3.7), in all aspects of confined space entry, including proper use of entry equipment and emergency rescue procedures.

A written Confined Space Entry Program is only as good as the personnel who utilize it. Confined space entry mishap investigation reports traditionally point to a lack of proper training as the cause of fatalities in confined spaces. While it is true that training is key to safe confined space entry, it is also true that training for confined space entry is complex. It involves training in respirator use, first aid, lockout procedures, and use of safety equipment specifically identified for confined space entry. (Safety equipment used to make confined space entry is discussed in further detail in Chapter 6 of this text.)

Confined space entry training must be conducted for entrants, attendants, entry supervisors (competent qualified persons), and support personnel (e.g., emergency rescue team members). Because of the wide range of possible hazards involved in confined space entry in the wastewater industry, each plant must develop a training program tailored to its specific needs.

A major point to keep in mind is that while OSHA allows the employer to tailor training to his/her requirements, OSHA is very specific about how confined space training is to be administered. For example, the employer must ensure that training is provided before an employee is assigned any

– GENERAL USE –
CONFINED SPACE ENTRY PERMIT
HAMPTON ROADS SANITATION DISTRICT

WORK CENTER: _____ **LOCATION:** _____

DATE/TIME: _____
(Permit expires in 12 hours.)

PURPOSE OF ENTRY: _____

PERSONNEL INVOLVED:

Qualified Person _____   Entrants _____

Attendant _____   _____

                                                                   _____

                                                              Hot Work _____

                                                              Fire Watch _____

                                                              Rescue Team _____

FIGURE 3.6. Confined space entry permit.

49

EMERGENCY PROCEDURES

If an emergency should occur - first summon help. Do not enter a confined space until help arrives and entry can be made safely. If a person is down for no apparent cause, you must assume that toxic gases or oxygen deficiency could exist - do not enter without full protective gear and self-contained breathing devices.

PERMIT ISSUANCE

The confined space entry permit, when properly authorized, allows the person to whom it is issued to enter the area specified. The work shall not be started until the checklist on the reverse side is completed and all applicable items are marked "YES", all requirements met, any discrepancies corrected and the indicated signature has been obtained. The permit shall be retained in the facility file for one year.

ATMOSPHERIC TEST RESULTS

Parameter	PEL	Initial Test	Prior to Entry	2	3	4	5	6	7	8	9	10	11
								Hourly Test Results					
O_2	19.5% to 23%	___	___	___	___	___	___	___	___	___	___	___	___
Flam	<10% LEL	___	___	___	___	___	___	___	___	___	___	___	___
H_2S	<20 ppm	___	___	___	___	___	___	___	___	___	___	___	___
Other	___												

Signed
Qualified Person _____

FIGURE 3.6 (continued). Confined space entry permit.

ENTRY CHECKLIST

	YES	NO	NOT APPLICABLE
1. Procedure provided, reviewed and enforced?			
a. All procedures understood? Training complete?	___	___	___
b. Person on site at all times to enforce all procedures?	___	___	___
2. Confined space preparation.			
a. Power sources "OFF"? Locked and tagged out?	___	___	___
b. Rotating equipment locked out, removed or disconnected?	___	___	___
c. Contents removed and space flushed?	___	___	___
d. All compressed gas cylinders are removed from space? (Cylinders part of self-contained breathing apparatus do not apply)	___	___	___
3. Confined space preparation for potential engulfment.			
a. Can the lines carrying liquid or particulate matter be disconnected, blinded or double blocked and bled? YES_____ NO_____ N/A_____			
b. If YES, has this been done?	___	___	___
c. If NO, complete the following?			
i. Attendant assigned?	___	___	___
ii. Rescue team available?	___	___	___
iii. Rescue equipment available?	___	___	___
4. Confined space preparation for potential hazardous atmosphere.			
a. Can the lines carrying flammable, injurious or incapacitating substances be disconnected or blinded? YES_____ NO_____ N/A_____			
b. If YES, has this been done?	___	___	___
c. If NO, complete the rest of the permit?	___	___	___
5. Initial atmospheric testing performed?			
a. Instrument has been properly calibrated within one week and field tested?	___	___	___
b. Are all of the initial atmospheric test results within the ranges below? OXYGEN 19.5% TO 23%	___	___	___

FIGURE 3.6 (continued). Confined space entry permit.

51

FLAMMABLES   0% LEL
H₂S   0 ppm
OTHER   0 ppm
YES_____ NO_____ N/A_____

c. If YES, and 4.b has been done then entry can be authorized.
d. If YES, and 4.b has NOT been done then complete the rest of the permit.
e. If NO, complete the rest of the permit.

6. Precautions taken for a potential for hazardous atmosphere or hot work?
    a. Attendant assigned or continuous monitoring equipment in place.   ___ ___ ___
    b. Special equipment/tools needed are available?   ___ ___ ___
    c. All sources of ignition have been removed?   ___ ___ ___
    d. Are all of the initial atmospheric test results within the ranges below? ___ ___ ___
        OXYGEN   19.5% TO 23%
        FLAMMABLES   <10% LEL
        H₂S   <20 ppm
        OTHER   < PEL
        YES_____ NO_____ N/A_____
    e. If YES, then proceed to Section 8. (All items of 7. will be N/A.)
    f. If NO, complete the rest of the permit.

7. Precautions taken for a hazardous atmosphere?
    a. Rescue team available?   ___ ___ ___
    b. Rescue equipment available?   ___ ___ ___
    c. Self-contained breathing apparatus used?   ___ ___ ___
    d. Mechanically ventilated?   ___ ___ ___
    e. Flammable concentration is below 10% LEL? (If NO, DO NOT ENTER   ___ ___ ___
        SPACE.)

8. Atmospheric testing performed?
    a. Atmospheric testing prior to entry performed and documented   ___ ___
        on reverse side?
    b. Hourly atmospheric test results documented on reverse side if required? ___ ___
        (Hourly testing is NOT required if the entrant is wearing respiratory equipment)

FIGURE 3.6 (continued). Confined space entry permit.

52

- Always control hazards before entry is allowed.

- Always test the atmosphere before entry and hourly thereafter.

- Use entry permits correctly.

- Refer to the District's Confined Space Program for specific procedures.

- NEVER enter a confined space to assist an unconscious person; call the Fire Department.

- Test your emergency communications before you allow entry.

- You are the qualified person; other people's lives depend on your good judgement.

Hampton Roads Sanitation District

Safety Training Certificate

has successfully completed

Confined Space Entry Training

to be designated as a

Qualified Person

CERTIFICATION EXPIRES LAST DAY OF _____

FIGURE 3.7. Confined space training certification card. Another card, similar to the one shown here is provided to all workers who have received general confined space training.

confined space duty. Moreover, the employer must ensure that confined space entry personnel are informed if any duties have been amended and whenever there is a change in operations affecting the confined space.

Written certification of confined space training is required. The employer must ensure that the employee's name, date of training, and the signature of the trainer are recorded. The certification of confined space training must be made available for inspection by the regulators. Experience has shown that regulators tend to show up at the job site unannounced. When this happens, the employee's training record is usually not available to show the regulator. To alleviate this problem, you might have the workers carry a billfold-sized card like the one shown in Figure 3.7. More will be said later in this text about the importance of documenting safety training.

RESPIRATORY PROTECTION (29 CFR 1910.134)

Before discussing respiratory protection, a note of *caution* is provided: *NOTE:* Before a worker is given or issued a respirator that is to be worn on the job, the employer must properly protect the worker by training him/her in proper respirator use. Such training must include the type of respirator that is to be worn for the particular hazard to be guarded against. Additionally, each worker must know the respirator's limitations and the maintenance and cleaning requirements for proper respirator use.

It may not always be obvious to workers when a respirator is required. Whenever workers must enter a confined space or other vessel for maintenance, entry should not be made until the atmosphere (air) within the confined space or vessel is tested for flammable agents, oxygen content, and toxic agents (e.g., hydrogen sulfide, methane, carbon monoxide, and other toxic agents).

Lack of oxygen is the most common cause of deaths in confined space entry. As an example of what can happen, consider that in order to prevent fire or explosion, fuel storage tanks that have contained flammable materials are frequently inerted or purged with various gases such as nitrogen prior to allowing personnel to enter the space to perform required maintenance. Purging the tanks will prevent fires from welding or spark-making activities, but if a worker should enter this space without a proper respirator, he/she will quickly be overcome from lack of oxygen.

Normal breathing air contains about 21% oxygen, by volume. A typical worker's total lung volume is about 5.5 liters. During normal breathing, each inspiration and expiration involves about 500 ml of air. Of this 500 ml, about 140–160 ml occupies the tracheobronchial tree, where no interchange of oxygen takes place with the blood. Therefore, only 340–350 ml of air is actually exchanged in each inhalation.

The air sack (alveolar) air, that is, the air from deep within the lungs, that is exhaled contains only about 11–12% oxygen, but combined with what remained in the tracheobronchial tree, the net exhaled composition is about 16%. When the concentration of air being inhaled drops below 16% oxygen, symptoms of distress will occur. Loss of consciousness can happen at oxygen levels below 11%. Breathing will cease if the oxygen concentration falls below 6–7%.

OSHA requires employers in occupational settings to establish and administer an effective written Respiratory Protection Program. This requirement is vital when you take into consideration that the most common route of entry of chemicals and toxic substances into the body is by inhalation. Respirators are frequently used in the wastewater treatment and collection industry. If your facility requires the use of respirators, it must have a written Respiratory Protection Program and abide by the OSHA requirements for respirator use.

Whether for confined space entry, for change-out of chlorine cylinders or tank cars, or for sandblasting, coating or other operations, a Respiratory Protection Program is required. Along with a written program, OSHA also requires that a workplace assessment be made to determine which respiratory hazards are present and/or if the potential for respiratory hazards exists. From this determination, the employer must provide the correct type of respirator to protect workers. The workplace assessment should be conducted by a knowledgeable person who is familiar with the workplace, the working conditions, and the workplace hazards.

There are three basic kinds of respirators: Air-purifying, supplied-air, and Self-Contained Breathing Apparatus (SCBA). *Air-purifying respirators* clean contaminated air before it is inhaled. There are different types of air-purifying respirators. One type removes particles from the inhaled air and is designed to filter out dusts, mists, sprays, and fumes. The other type, the gas- and vapor-type air-purifying respirator, is designed to absorb or chemically remove gases and vapors.

Air-purifying respirators have limitations. Probably the most significant limitation is the fact that they *do not* provide a supply of respirable air. Secondly, air-purifying respirators do not protect the user from all types of hazards; not all dangerous substances can be safely filtered out of the atmosphere. A third limitation is the filtering media. Filters can and do become clogged and have a limited lifespan.

In some cases it is more desirable to use the second type of respirator, the *supplied-air respirator,* which supplies air from an outside source. Supplied-air respirators are used whenever there is not enough oxygen in a situation or area where the concentration that is present is Immediately Dangerous to Life or Health (IDLH). An IDLH atmosphere is defined as a very hazardous atmosphere where exposure can cause serious injury or death in a short period of time.

Two kinds of supplied-air respirators are available: air line-type supplied-air and hose mask type. The *air line-type supplied-air respirator* supplies clean, compressed air from a stationary source delivered through a pressurized hose. The *hose mask type* supplies clean, noncompressed air through a strong hose to a respiratory inlet covering. Air may be supplied with a motor or a hand-operated blower.

The supplied-air respirator has limitations. The biggest disadvantage is that the loss of the source of respirable air supplied to the respirator inlet covering voids any protection to the wearer. Thus, adequate protection may not be provided in certain highly toxic environments. The air intake of the device used must be fed uncontaminated air, which might not be available. Moreover, mobility may be restricted by the inconvenience and length of the air supply hose (which cannot exceed 300 feet).

When mobility is an issue and when respiratory protection against all breathing hazards is required, the *self-contained breathing apparatus (SCBA)* type respirator is required. Although the SCBA allows the user more freedom and protection than the other types of respirators, it is still important to monitor the amount of air used and the amount available for use in the worker's air cylinder.

Selecting the correct type of respirator to wear is only one of several requirements involved with respiratory protection. Because it does little good (and could be disastrous) to have workers wear respirators if the respirator is not suited for the hazard they are to be exposed to, OSHA requires the employer to ensure that the proper respirator is selected. For example, if an employee chooses to wear a 1/2 face mask, air-purifying type respirator in a confined space that has an atmosphere composed of 100% carbon dioxide, then the worker will become another confined space fatality. Thus, employers must ensure that they provide their workers with the correct type of respirator to protect them from the respiratory hazards they are or might be exposed to.

Choosing the right respirator is not always easy. To help you in the selection process, Figure 3.8 is provided. The flow chart depicted in Figure 3.8 provides guidelines for selecting appropriate respirators. Supervisors and workers need to be aware of this flow chart. More importantly, they need to know how to use it.

After using Figure 3.8 in the process of selecting the correct respirator to protect against the identified hazard, and if you have determined that a cartridge-type air-purifying respirator will suffice, you need to ensure that the respirator is equipped with the correct cartridges. Information on the correct cartridges to be used can be obtained from the respirator manufacturer or a reputable respirator sales representative.

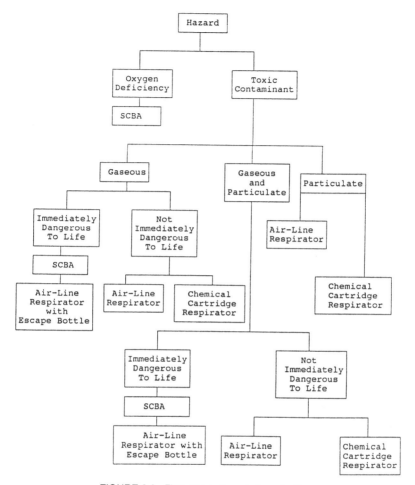

FIGURE 3.8. Flow chart of respirator selection.

Wearing the correct respirator is important, but making sure that the respirator actually fits is just as important. OSHA requires that workers who are designated to wear respirators in the workplace receive respirator fit tests. The purpose of respirator fit-testing is to provide the worker with a face seal on the respirator that exhibits the most protective and comfortable fit. There are two types of respirator fit tests: *quantitative* and *qualitative*. In both tests, a harmless smoke, gas, vapor, or aerosol is used to test for fit.

In a quantitative fit test the person to be tested is instructed to don his/her respirator and is then placed inside an enclosure where he/she is exposed to a test agent (irritative smoke, banana oil, or saccharin). During the quantitative fit test the worker is carefully monitored and observed to determine if any of the test agent is entering the mask. Special instrumentation is used, and the test is generally performed by an industrial hygienist or a safety professional who has extensive training in this area. The special instrumentation and equipment required to perform quantitative fit tests are expensive.

In the qualitative test (this is the test that is most commonly administered in general industry), the person performing the fit test knows the worker has a good fit if the worker cannot sense the agent when it is released in close proximity to the worker's breathing zone (between the shoulders and the top of the head). The qualitative fit test is not as accurate as the quantitative fit test because it relies on the wearer's subjective response, which may not be entirely reliable. However, the qualitative fit test is much easier to administer, less expensive, and easily performed in the field.

At a minimum, before using a respirator, the user must test a respirator by performing his/her own pre-use self-fit test. Each time the worker dons a respirator, he/she must test the mask for positive and negative seal. This is accomplished by performing positive and negative pressure tests.

During the *positive pressure test,* the user closes the respirator's exhalation valve and breathes out gently into the face piece (see Figure 3.9). This may cause the face piece to bulge out slightly. If no air leaks out, the user has a good fit.

During the *negative pressure test,* the user covers both filter cartridges with the palms and inhales slightly to partially collapse the mask. This negative pressure is held for 10 seconds. If no air leaks into the mask, it can be assumed the mask is fitting properly (see Figure 3.10).

In order to retain their original effectiveness, respirators must be cleaned (disinfected) and properly stored. The provisions for cleaning and proper storage must be included in the organization's written Respiratory Protection Program. Respirators should be cleaned after each use. When performing the cleaning operation, it is best to follow the manufacturer's instructions. Caution must be exercised whenever respirators are cleaned. Certain detergents and solvents will not only damage the face pieces but could also damage the respirator casing.

After using and cleaning a respirator, it should be properly stored. For example, respirators are subject to damage when stored in bright sunlight. Moreover, they can become contaminated or damaged when exposed to dirt, grease, oil, solvents, or other contaminants. This can render the respirator useless and create a serious health hazard for the user.

In terms of the user's health, OSHA states that not everyone can wear a respirator. Anyone assigned to a job requiring a respirator must first undergo a medical examination and be cleared by a medical doctor for

FIGURE 3.9. Fit test–self-test—positive pressure.

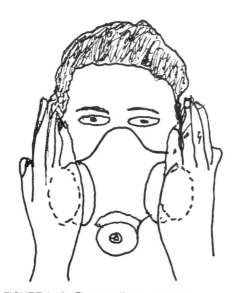

FIGURE 3.10. Fit test–self-test—negative pressure.

respirator use. OSHA is very straightforward on this issue, stating that no one should be assigned a task requiring use of respirators unless found physically able to do the work while wearing the respirator. Follow-up medical examinations are also required — usually at no longer than five-year intervals, depending on the age of the worker. If the worker is forty or older, annual medical examinations should be required. The ultimate decision on what the actual frequency of reexamination is to be made by a medical physician.

The most widespread medical screening procedure in pre-placement and medical surveillance testing of a worker's medical suitability to wear respirators is called a *pulmonary function test* or *spirometry* (see Figure 3.11). The pulmonary function test measures two important parameters: *FVC* and *FEV1* (see Figure 3.11).

FVC, or Forced Vital Capacity, measures the worker's vital capacity when it is exhaled as rapidly and forcefully as possible. The vital capacity (lung volume) is the amount of air that can be exhaled after a maximal inspiration. FVC decreases with the loss of lung volume that occurs in restrictive lung diseases resulting from exposure to various fibrogenic dusts such as asbestos and silica.

FEV1, or Forced Expiratory Volume–1 Second, measures the amount of air the worker is able to expire in the 1st second. FEV1 divided by FVC (FEV1/FVC) decreases in obstructive disease and is caused by cigarettes, asthma, exposure to chemicals, and other conditions.

American National Standards Institute (1984) recommends that FVC measure 80% or greater and FEV1 70% or greater. If FVC is less than 80% or the FEV1 is less than 70%, restriction from respirator use should be considered. However, the ultimate decision on whether or not the worker is disqualified from respirator use for medical reasons must be made by competent medical authority. The safety practitioner must keep in mind that although the pulmonary function test is an excellent screening tool, it is not the definitive test or statement as to whether one is medically fit or not; again, only competent medical authority can make this decision. Pulmonary function testing can also provide the employer with baseline data against which an assessment can be made of any physiological changes in respirator wearers that might occur with the passage of time.

Before the organizational safety official performs pulmonary function tests on employees, two requirements must be met. First, the safety official must be formally trained as a NIOSH Pulmonary Function Technician. This training usually can be completed in no more than three days at a NIOSH-approved training center. Training as a pulmonary function technician does not require a medical degree or background. The second requirement involves equipment. Spirometers are available from various vendors and usually can be purchased for less than $3,000.

```
PATIENT DATA and TEST CONDITIONS

Name  :  _____
ID No.:        1
Age   :        51  years
Sex   :        male
Race  :        caucasian
Height:        68  in.

Tech. :  _____
Date  :        08/03/95
Time  :        03:20
Ambient Conditions:  760 mm Hg at 22.2 C
Selected Normals  :  Knudson

Doctor:  _____

_____

Best Test Summary  Tests 01-01  08/03/95

Index       No    Meas    Pred   %Pred   %Var

FVC         01    5.60    4.28   131     n/a
FEV 1       01    4.37    3.48   126     n/a
FEV1/FVC%   --    78.0    81.3   96      n/a
FEF25-75    01     3.9     3.6   108     n/a
PEF         01    10.0    n/a    n/a     n/a

Spirometric results appear normal
```

FIGURE 3.11. Pulmonary function test. A sample computer printout of pulmonary function test results is also shown.

61

A final word on pulmonary function testing. Experience has shown that workers generally look forward to this test. This is especially the case for those employees who are active cigarette smokers and those who have recently given up smoking. When a worker receives a pictorial graph and/or accompanying parameters that roughly point out what is going on inside his/her lungs, he/she is interested. It can be very gratifying to observe workers who give up cigarette smoking as a direct result of their annual pulmonary function tests. Persuading workers to give up cigarettes not only helps workers maintain their good health, it also helps significantly reduce their sick leave requests.

Breathing problems, such as asthma, allergies, or emphysema, make breathing difficult. When a worker suffers from one of these maladies and is required to wear a respirator and then performs ardous work in a confined space, the worker's well-being is put at risk.

In addition to these medical problems, workers should be screened for other potential respirator use limitations. For example, if a worker suffers from claustrophobia, he/she should not be required to wear a respirator. Moreover, workers who suffer from high blood pressure or heart disorders could be put under great stress if forced to wear respiratory protection. Finally, although rare, a worker might have facial scars or other major facial abnormalities that would not permit that person to wear a respirator on the job.

A viable and effective Respiratory Protection Program is only achievable if proper training on this vital topic is administered. Since the very real possibility exists for the worker who is required to wear a respirator on the job, to perform his/her job-related function in a life-threatening environment, proper training in this area cannot be overemphasized. Again, the training must be documented.

Documentation is also required for respirator use. Respirators must be inspected prior to and after each use. A record of these inspections is required. When OSHA audits an organization's Respiratory Protection Program, it *always* asks to review each work center's Respirator Inspection Records. In addition to the prior-to-use and after-use record of inspection, respirators must be inspected on a weekly and monthly basis. Again, these weekly and monthly inspections must be recorded in the work center respirator recordkeeping forms. During your inspection of respirators, if any discrepancy is discovered and/or repair is made, these actions must also be recorded in the Respirator Inspection Record.

For those desiring more information about OSHA's Respiratory Protection Program, several important publications are available. One such publication is NIOSH's *Guide to Industrial Respiratory Protection*. This publication is the ultimate single source of respirator protection information and can be obtained from the U.S. Department of Health and Human Services.

HEARING CONSERVATION (29 CFR 1910.95)

A potential hazard of the workplace that is often overlooked is excessive noise. Noise is usually defined as any unwanted sound. This sound is caused by rapid fluctuations of air pressure on the listener's eardrum. Sound may be unwanted for several reasons: it causes hearing loss, has adverse effects on human physiology, interferes with normal conversation, or is just plain nuisance. Sound is measured in decibels (dB). The decibel is a dimensionless unit used to express physical intensity or sound pressure levels. If sound is intensified by 10 dB, it seems to human hearing approximately as if the intensity has doubled. Conversely, reduction by 10 dB makes it seem as if the intensity has been reduced by half.

The reference point for noise level measurement is 0 dBA, which is the threshold of hearing for a young child with very good hearing. The threshold of pain for humans is 120–125 dBA. The "A" part of dBA is determined by using instruments used to measure sound levels. These instruments have a measuring scale designed to resemble sound to which the human ear (A-weighted sound level) is sensitive. As illustrated in Figure 3.12, noise is everywhere.

As mentioned, noise is a workplace problem in many instances. Along with possibly causing permanent or temporary hearing loss, noise also affects the nervous system. Anyone who has worked at a wastewater treatment plant or pumping station knows that the wastewater treatment process produces noise: from the low, barely audible hum of fractional horsepower electric motors to the din created by 300 + horsepower aeration blowers and diesel engines, the noise intensity spectrum at a wastewater treatment facility and/or pumping station is wide.

Hearing loss is a common workplace injury that is often ignored. It was not uncommon in the past for workers to accept partial hearing loss as a cost of working in a noisy workplace, including a wastewater treatment plant. Times have changed, however. OSHA recognizes noise for what it really is: an occupational hazard that can cause temporary or permanent hearing loss, stress, and other physical problems.

Because noise is a workplace hazard, OSHA has established criteria for protecting workers' hearing. The main factors related to hearing loss are intensity, time of duration of exposure, and repeated impact noise. It is the continuous exposure to high-level noise that must be avoided. With increased time of exposure, there is a corresponding increase in harm done.

In order to protect workers, and to abide by OSHA requirements, waste-

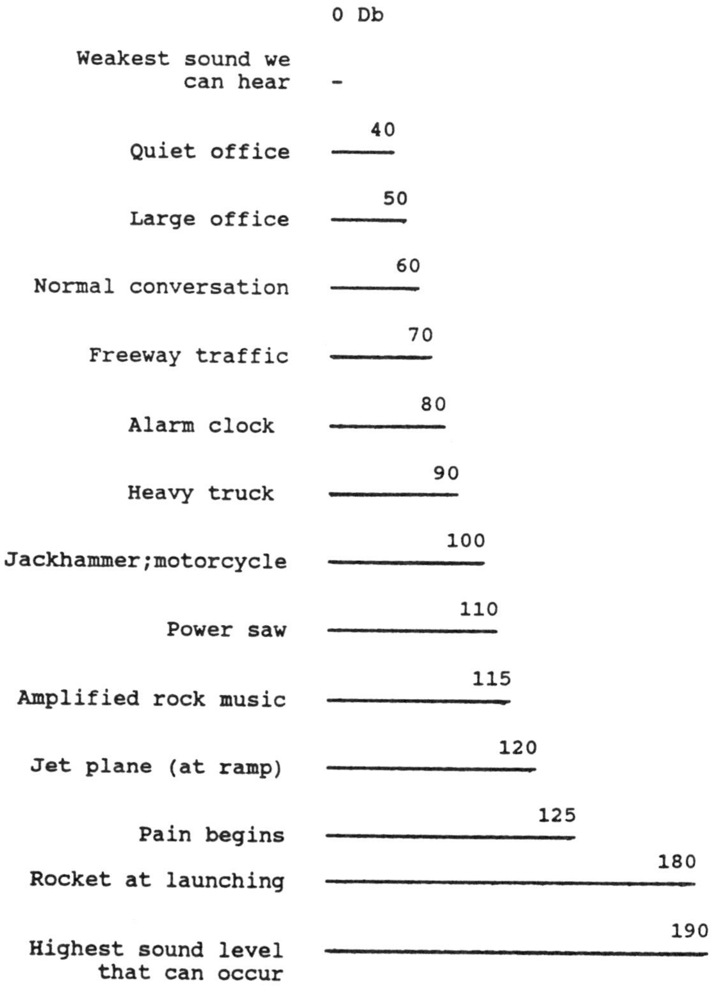

```
                          0 Db

Weakest sound we
      can hear          -

                          40
Quiet office           ———————

                          50
Large office           ———————

                          60
Normal conversation    ———————

                          70
Freeway traffic        ———————

                          80
Alarm clock            ———————

                          90
Heavy truck            ———————

                          100
Jackhammer;motorcycle  ———————

                          110
Power saw              ——————————

                          115
Amplified rock music   ——————————

                          120
Jet plane (at ramp)    ———————————

                          125
Pain begins            ————————————

                                        180
Rocket at launching    ————————————————————

                                          190
Highest sound level    ——————————————————————
      that can occur
```

FIGURE 3.12. Noise levels (*Source:* OSHA).

water facilities need to incorporate a Hearing Conservation Program into their overall safety program whenever sounds generated in the workplace are irritating to workers. For example, when a worker needs to raise his/her voice to shout to be heard by someone closer than one foot away, the worker is being exposed to noise levels that are too high. Moreover, whenever measured sound levels reach 85 decibels or higher for an 8-hour time period, a Hearing Conservation Program is required by OSHA.

When beginning to implement a Hearing Conservation Program, the plant safety official usually asks the same standard question: *Where do I start?*

The first step is to conduct a noise survey using a calibrated sound meter of the plant site, including pumping stations and maintenance centers. Each building at each facility should be surveyed for sound. Keep in mind that even administrative office areas, at times, can contain high noise levels.

When conducting the plant sound survey, it is important to remember that various machines might not be on line and in operation at the particular moment the survey is conducted. It is impossible to conduct a sound survey that will provide a true picture of actual noise levels unless all machinery is running.

Before conducting the plant sound survey, it is wise to draw a rough map of the layout of each building. On this map you should include pictorial representation of each noise-making machine or device. When testing for sound level, start at the machine and work around it; it is important to survey the entire area around the machine. Moreover, the sound survey should be conducted at varying distances from each machine. Sound measured at a certain distance from the source in open air is reduced by about 6 dB for each doubling of that distance. Sound is reduced less when it spreads inside a room. The point is that by moving away from the sound source, the sound level is reduced (see Figure 3.13).

When measuring for noise level, measure the sound level at a pump or motor or blower starting at a close-in distance (within 2 feet), and then move outward by doubling the distance from the machine: to 4 feet, 8 feet, and 16 feet, respectively. On the rough sketch, at the particular machine being measured, draw concentric circles at each distance and record the meter readings on each circle.

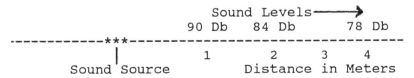

FIGURE 3.13. Noise and distance relationship (*Source:* OSHA).

The concentric circle approach aids in two significant ways. First, by knowing the sound level for varying distances from the noise source, it is possible to label or mark off high-noise areas and direct routine traffic away from the noise. Second, this approach provides a map of where the noise hazards are. The written program should require labeling or marking of these areas (see Figure 3.14).

```
                         DANGER!
             HEARING PROTECTION REQUIRED IF TIME
         WORKING IN THIS SPACE EXCEEDS _____MINS/HRS.

NOTE: In order to determine the time
      exposure to record in the MINS/HRS
      block above, the following is to be used.

             Permissible Noise Exposures
```

SOUND LEVEL in DECIBELS	MAXIMUM TIME OF EXPOSURE
85--Program required	8 hours
90--Hearing Protection required	8
92	6
95	4
97	3
100	2
102	1.5
105	1
110	0.5
115	0.25

CAUTION

HEARING PROTECTION REQUIRED

FIGURE 3.14. Hearing conservation warning signs (*Source:* OSHA).

When it has been determined what type of noise hazard warning will be used and where each sign is to be posted, the next item that should be included in the Hearing Conservation Program is an evaluation of engineering and administrative controls.

The possibility of incorporating engineering and administrative controls into your Hearing Conservation Program should be looked at closely. In the world of safety it is always best, if possible and feasible, to "engineer out" a safety hazard instead of guarding against it. Engineering out the noise problem might include installing mufflers on air exhaust nozzles or isolating a machine. If the noise source cannot be damped, maybe something can be done about the noise path. Placing sound-absorbent enclosures around noisy equipment, for example, might be another modification that can be easily accomplished. As a last engineering control measure, protecting the receiver of the noise is a solution. This can be accomplished by constructing an enclosure around the employee's work station.

When engineering controls are not possible, feasible, or cost-effective, administrative controls are called for. The primary administrative control used to protect workers from unwanted sound or noise is to issue each worker hearing protection devices.

Hearing protection devices include earplugs and earmuffs. Properly installed, foam earplugs offer the best protection. Yet, many workers feel that earmuffs, which fit over the outside of the ear, provide more protection than earplugs. This is usually not the case, however. Earmuffs are only as good as the seal they provide around the ear. In extremely noisy environments it might be wise to recommend *double hearing protection*: wearing ear plugs and earmuffs at the same time.

Another important area that needs to be addressed in the plant Hearing Conservation Program is the designation of *audiometric evaluation* procedures. OSHA requires employers with facilities where noise exposure equals or exceeds an average of 85 decibels or more over an 8-hour period to provide their employees with audiometric testing.

Wastewater treatment plant safety officials can ensure that audiometric examinations are conducted by either performing the tests themselves or by hiring a local medical contractor to perform the tests. If tests are conducted in-house, the audiometric technician must be properly trained. NIOSH-approved training centers offer audiometric technician or occupational hearing conservationist training throughout the country. Such training can usually be completed in three days and the trainee is not required to have a medical background. Moreover, audiometric equipment may be purchased at reasonable prices. The plant safety official must keep in mind that audiometric testing must be conducted in a "noise-free" environment (e.g., in an audiometric testing booth).

Whatever audiometric testing approach the plant safety official decides to implement, in-house testing or hiring a medical contractor, the audiometric testing procedure is subject to certain mandatory requirements. For example, each employee who might be exposed to noise levels at or greater than 85 decibels must be given a baseline audiometric test. The audiometric test should be conducted only when the worker has not been exposed to workplace noise for at least 14 hours. Testing must be performed by a certified audiometric technician, using a calibrated audiometer (annual requirement) in an environment of less than forty dB(A). Six months after receiving the baseline audiometric test, the worker should be tested again—then annually thereafter. All new employees should be given the baseline audiometric test within six months of hire date. Any time a worker experiences a Temporary or Permanent Threshold Shift, he/she should be retested within 60 days.

Temporary Threshold Shift (TTS) is known as auditory fatigue, which is caused by excessive exposure to noise. TTS is transient, but its presence is undesirable. *Permanent Threshold Shift* (PTS), on the other hand, is a more serious hearing disability. As the name implies, PTS can result in permanent loss of hearing acuity and should be rigorously guarded against.

If the follow-up audiometric test suggests Permanent rather than Temporary Threshold Shift, the worker must be referred to a physician for evaluation to determine whether or not damage is permanent. All cases of occupational-related hearing loss must be recorded on the OSHA-200 Log. The plant's Hearing Conservation Program must receive oversight guidance from a licensed audiologist.

Because hearing loss is such a gradual process, it is often ignored by the worker and the supervisor. This is why training is so critical. Like all safety programs, the programs are only as good as the training that accompanies them. In a Hearing Conservation Program, not only is training critical, so is supervision and management. An effective Hearing Conservation Program cannot be established without a written program, worker training, and effective supervisory management.

It has often been said that leaders should lead by example. That is, the worker will not wear the hearing protection devices unless the leader leads the way by wearing hearing protection devices and ensuring that the workers also wear them. Hearing loss liability continues to be a costly proposition for many high-noise industries. By incorporating an effective hearing conservation program into the organizational safety program, you will reduce worker hearing loss and liability for the loss.

PERSONAL PROTECTIVE EQUIPMENT (PPE)
(29 CFR 1910.132 AND .138)

LaBar (1990) points out that according to an Eastern Research Group survey, appropriate use of Personal Protective Equipment (PPE) could have prevented as much as 37% of the on-the-job injuries and illnesses that were reported. OSHA understands the importance of PPE. On October 5, 1994, its revised Personal Protective Equipment (PPE) Standard went into effect in workplaces throughout the United States. It is important to note that OSHA stresses that PPE should *not* be used as a substitute for engineering, work practice, and/or administrative controls. Instead, PPE should be used in conjunction with these controls to provide for worker safety and health in the workplace; that is, as backup or secondary protection.

The PPE addressed in the new OSHA Standard specifically deals with protective equipment designed to protect many parts of workers' bodies including eyes, face, head, hands, and feet. PPE for Respiratory Protection and Hearing Conservation are addressed in other OSHA standards and have already been discussed in this text.

The part of the PPE regulation that applies to a particular organization varies, depending primarily on the *types of hazards* present in the workplace environment. The PPE regulation is known as an OSHA "Performance Standard." A Performance Standard mandates that protection be provided by the employer to the employee, but the employer is allowed to meet the minimum requirements as determined by experience/performance and specified by the organizational safety official.

In general, OSHA's PPE Standard covers three important areas:

- hazard assessment of the workplace and certification
- selection of PPE
- worker PPE training and certification

Before addressing the individual areas of PPE, it must be reiterated that OSHA's goal is to use PPE *in conjunction with* other controls to protect workers. PPE is simply designed to create a barrier between the worker and workplace hazards; *it does not remove the hazard from the workplace.*

The first step in determining the types of PPE that are needed at the plant site is to conduct a hazard assessment survey to determine if hazards that require the use of PPE are present or are likely to be present. If the plantwide *hazard assessment* survey shows that hazards, or the likelihood of hazards, are possible, employers must select and have affected employees use *properly fitted* PPE, suitable for protection from the existing hazards. In addition, standby PPE protection should be on hand in case potential hazards become actual hazards.

In order for PPE to be effective, it must be properly selected. To aid in the selection process, each employer is required to perform the workplace hazard assessment mentioned earlier. As part of this process, a written *Certification of Hazard Assessment* must be prepared, which at a minimum contains the following information:

- location of the workplace evaluated
- details of the hazards assessed
- the person certifying the evaluation
- dates of hazard assessments

Figure 3.15 shows a certification of hazard assessment form. OSHA not only requires that such a form or format be used in making the PPE assessment, it also requires that the assessment be in writing and signed by the person who performed it. It is important for the plant safety official to retain a copy of the PPE assessment and have it readily available for review by regulators whenever they arrive at the plant site.

Once the workplace hazard assessment has been completed, the employer is required to *select and require* each employee to use the correct PPE. In order for PPE to be effective in hazard control, workers must be aware of the importance of their role in the proper fitting and wearing of it. Moreover, the appropriate PPE for each workplace must be *clearly described* to each worker in that area. Work areas where hazardous operations such as welding and chemical handling take place must be labeled for the hazard and for the required PPE.

According to the Bureau of Labor Statistics' injury reports, a significant number of workers who are injured on the job have never been trained in the proper use of PPE. Worker training in the proper wearing and use of PPE is essential. Workers must be trained to know when to wear PPE; they must know what PPE to wear; they must know how to wear the PPE; they must know the limitations of PPE; and, they must be trained in the proper care, maintenance, useful life and disposal of PPE. Once the worker has demonstrated an understanding of PPE training, the safety official must verify the worker's certification in writing.

By now you have picked up on the fact that the author feels that the importance of training cannot be overemphasized. As important as training is, however, it is just as important to document that training has been completed. Training completed but not properly documented is in reality training that was never done; that is, if you cannot demonstrate to OSHA in writing that training was accomplished, then, in the regulator's view, it was not.

To verify the effectiveness of PPE training or any other safety training, it is always prudent to give a test or quiz at the end of each presentation. Such a test or quiz should be straightforward and directly on the subject matter that was taught. The test should be written in such a manner that all workers will be able easily to understand its content. The test should be reviewed and all questions that were incorrectly answered should be answered correctly. This ensures that the worker who may have answered any question

HAMPTON ROADS SANITATION DISTRICT
PPE HAZARD ASSESSMENT FORM

JOB CLASSIFICATION:

LOCATION OF JOB, FUNCTION, ACTIVITY OR SITUATION:

APPARENT HAZARDS:

     IMPACT:

     PENETRATION:

     COMPRESSION (ROLL-OVER)

     CHEMICAL:

     HEAT:

     HARMFUL DUST:

     LIGHT:

     NOISE LEVELS:

     RESPIRATORY:

PERSONAL PROTECTIVE EQUIPMENT NOW PROVIDED FOR THIS JOB FUNCTION:

ADDITIONAL PERSONAL PROTECTIVE EQUIPMENT THE IS NEEDED FOR SAFE JOB PERFORMANCE:

PERSONAL PROTECTIVE EQUIPMENT TRAINING:

PERSONAL PROTECTIVE EQUIPMENT MAINTENANCE, FUNCTION, SPECIFICATIONS, DUTIES:

ASSESSMENT COMPLETED BY:

SIGNATURE: DATE:

TITLE:

FIGURE 3.15. PPE hazard assessment form (*Source:* Hampton Roads Sanitation District; used by permission).

incorrectly does not leave the training area without knowing the correct answer. Workers who have difficulty with reading and writing should be administered oral tests/quizzes. The safety official should use discretion whenever this practice is followed. The idea is to test the worker's comprehension of the subject matter, not to embarrass the worker.

Upon successful completion of PPE or any other safety training, it is a good idea to present each worker with some type of certificate of completion. According to OSHA regulations, with regard to PPE training, each employee must receive a written *Certificate of Employee Training* to verify that he/she has received and understands the required information. This certificate must contain the employee's name, the date of his/her training, and the subject of certification.

Figure 3.16 shows a training certificate used at HRSD. In this age of computer technology, it is rather simple to generate in-house training certificates. The workcenter supervisor might want to make copies of each certificate and place the copies in the workcenter training record as well as the worker's personnel record. Remember, training that is not documented is

FIGURE 3.16. Training certificate (*Source:* Hampton Roads Sanitation District, used by permission).

ATTENDANCE ROSTER

SUBJECT MATTER

In accordance with the recordkeeping and training requirements of the Personal Protective Equipment (PPE) Standard, OSHA 29 CFR 1910. 132 and 1910.138, I have received training on when PPE is required, what PPE I must wear, how to wear the PPE, the limitations of PPE, and the proper care, maintenance, useful life and disposal of PPE. I have agreed to verify my understanding and training of 1910.132 and 1910.138 by signing and dating this form.

WORKER'S NAME: DATE:

- - - - - - - - - - - - -   - - - - -

- - - - - - - - - - - - -   - - - - -

- - - - - - - - - - - - -   - - - - -

- - - - - - - - - - - - -   - - - - -

- - - - - - - - - - - - -   - - - - -

- - - - - - - - - - - - -   - - - - -

- - - - - - - - - - - - -   - - - - -

FIGURE 3.17. Training attendance roster used to verify the worker's attendance at a particular safety training session and also serves as verification that the worker understood the training. This form or a similar form should be used for all safety training sessions and should be filed in the workplace master training file.

training that was never accomplished—this is the view OSHA and the courts might take.

In addition to testing and certifying workers who have completed safety training, it is prudent to have each worker sign and date an attendance roster. Figure 3.17 shows one version of such a roster.

As for the subject content of OSHA's PPE Standard, four distinct areas are addressed: eye and face, head, hands, and foot protection. The *eye and face protection* part of the standard requires that workers be provided with eye and face protection whenever they are required to work with:

- liquid chemicals
- hazardous gases
- flying particles
- molten metals
- injurious radiant energy

OSHA requires that eye protection devices meet the following minimum requirements:

- provide adequate protection against the particular hazard for which they are designed
- be reasonably comfortable when worn under the designated conditions
- fit snugly without interfering with the wearer's movements or vision
- be durable
- be capable of being disinfected
- be easily cleanable
- be kept clean and in good repair

Safety glasses are the basic form of eye protection. OSHA now requires protective eye coverage from the front and the sides any time there is a hazard from flying objects. Such coverage can be accomplished by using safety glasses with attached side shields or by using detachable side protectors.

There are several types of eye and face protection available (see Figure 3.18), including:

- safety glasses
- goggles

FIGURE 3.18. Types of eye and face protection.

- face shields
- full hoods
- welding helmets

A common mistake when workers wear eye protection is that most workers think that a face shield is eye protection. Face shields are designed to protect the worker's face; they are not eye protection. Safety glasses or goggles should be worn under face shields to provide primary eye protection.

Another common mistake occurs when workers use the wrong type of eye protection for a particular work assignment. As a case in point, consider workers who must work with corrosive chemicals. These workers normally wear safety goggles. However, the goggles may be the wrong type for the work being performed.

Safety goggles are available in vented, nonvented, and shielded-vent types. When working with liquid corrosive chemicals, the worker should wear goggles that are not vented or have shielded vents. When working with gaseous chemicals, the worker should wear nonvented goggles only. Vented goggles are to be used only for grinding operations or for assignments where chemicals are not used. In addition to requiring workers to wear the correct type of goggle, the plant safety official should also require workers to wear a face shield for added safety.

Many employees wear prescriptive eyewear on the job. When these workers are required to wear eye protection, they must also wear their approved eye protection. Some workers who wear prescriptive glasses wrongly assume that they are automatically wearing eye protection. This is not the case, however. Employers are not required to provide workers with prescription protective eyewear. If the worker desires to purchase his/her own protective eyewear that is ground to his/her prescription, the protective eyewear must meet the applicable ANSI (American National Standards Institute) standard. It is wise to issue workers who wear prescriptive eyewear with goggles that are oversized; that is, designed to fit over their regular prescription glasses. This seems to suit the workers' needs while also providing the required protection.

Head protection is required for workers who are or might be exposed to injury from falling objects or who work near exposed electrical conductors which could contact the head area. A government study on disabilities suggests that there are over 65,000 head/face injuries each year. Another study conducted by the Bureau of Labor Statistics of accidents and injuries noted that most workers who suffered impact injuries to the head were not wearing head protection. Although head protection would not have prevented all of these injuries, there are plenty of workers who have suffered head injuries who wish they had been better protected at the time of their injury.

The primary head protection recommended is the hard hat. Hard hats are designed to protect the worker from impact and penetration caused by objects hitting the head. They also provide limited protection against electrical shock. Hard hats are tested to withstand the impact of an 8-pound weight dropped 5 feet. They must also meet other requirements including weight, electrical insulating properties and flammability.

A 1980 Bureau of Labor Statistics survey found that 91% of injuries to persons wearing hard hats were struck other than on top of the head. Lateral head protection is available but it is larger, heavier, hotter, and more expensive than standard hard hats (Minter, 1990).

Hard hats come in three classes: A, B, and C. Class A are made from insulating material that is designed to protect the worker from falling objects and electric shock by voltages of up to 2,200 volts. Class B hard hats are made from insulating material designed to protect the worker from falling objects and electric shock by voltages up to 20,000 volts. Finally, Class C hard hats, while designed to protect workers from falling objects, are not designed for use around live electrical wires or where corrosive substances are present.

Most organizations require workers to wear the Class B hard hat. Since corrosives, overhead objects (cranes), and electrical circuitry are present in most wastewater treatment facilities, the Class B hard hat is also best suited to protect workers in this environment (see Figure 3.19).

It is important to label or post those areas in the plant site where hardhats are required (see Figure 3.20). When major construction work is in progress on the plant site, hardhats should always be worn by plant operators who might have to make hourly rounds in, through, or near these areas. Hard hats should be routinely inspected and replaced at regular intervals.

FIGURE 3.19. Class B hard hat.

CAUTION
HARD HAT
AREA

FIGURE 3.20. Hard hat warning sign.

Arm and hand on-the-job injuries are frequent among workers in wastewater treatment facilities. OSHA's PPE Standard requires employers to provide hand protection for workers for on-the-job use.

The key ingredient to proper finger, hand and arm protection is to ensure that workers wear protection when they are exposed to hazards such as those that occur due to skin absorption of harmful substances, severe abrasions, cuts, lacerations, punctures, chemical burns, thermal burns, vibration, and harmful temperature extremes.

Gloves are the most common type of hand protection. The type of glove selected depends upon the type of work involved. Regardless of the type of glove selected, it must fit the worker's hand. Other devices can be used to protect the worker's fingers. For example, finger cots are designed to protect individual fingers or fingertips. Thimbles protect the thumb. Further, the palm can be protected from cuts, friction, and burns by using hand pads. Long sleeves and forearm cuffs protect arms and wrists from heat, splashing liquids, impact, and cuts. When hand lotions and creams are used for hand and finger protection, they are not to be used as a substitute for gloves.

In the wastewater treatment plant laboratory or chemical handling area, special gloves should be used. These special gloves include nitrile or synthetic rubber gloves for handling oils, some solvents and grease. Neoprene-type gloves can be used for handling a broad range of chemicals, oils, acids, caustics and solvents. Finally, polyvinyl chlorine (PVC) gloves can be used in handling acids, caustics, alkalis, bases and alcohol.

When working with chemicals, the selected glove type should protect against the toxic properties of the chemical, not only for local effects on the skin but also for systemic effects. Generally, gloves that are rated "chemical resistant" can be used to work with dry powders. Whenever using gloves to protect against chemicals, it is important to remember that the gloves must be removable without contaminating the skin. When in doubt whether or not gloves are required for chemical handling and what type of glove should be used, consult the chemical Material Safety Data Sheet (MSDS) or ask the manufacturer.

Finally, it is important to provide workers with gloves that will protect them against severe temperature extremes and vibration. For example, when work is to be performed around or on plant incinerators, a special heat-insulating glove should be issued. In the winter when a worker is assigned to work on an excavation in the field, he/she should be issued gloves that protect against frostbite. Anti-vibration and impact gloves are available for use by workers who perform jackhammering and other high-vibration type jobs. The purpose of using impact and anti-vibration gloves is to dampen the shock before it reaches the worker's hands. Figure 3.21 shows a few examples of gloves that workers can wear to provide hand and finger protection from some hazards. *Note:* Whatever type glove is chosen to protect the worker's hands and fingers, the glove must remain functional in terms of dexterity, comfort, durability, and cost.

According to a Bureau of Labor Statistics survey, most workers in selected occupations who suffered foot injuries were not wearing protective footwear. In the wastewater industry, foot injuries usually occur for the same reason. Foot injuries usually occur when heavy or sharp objects fall (typically falling fewer than 4 feet) on the foot, when something rolls over

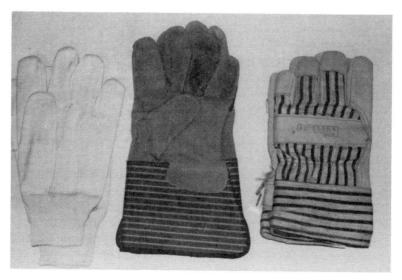

FIGURE 3.21. Safety gloves.

the foot, or when the worker steps on an object that pierces the sole of the shoe. As a result, *foot protection* is another requirement of OSHA's PPE Standard.

Selecting the proper protective footwear for workers depends on the type of work they will be performing. For construction work (excavations) or work involving rolling objects (e.g., rolling chlorine cylinders), safety shoes with impact-resistant toes should be worn. For work in or around liquid corrosive chemicals, rubber or synthetic footwear are appropriate. When a worker is involved in work that is around exposed electric wires, metal-free, non-conductive shoes should be worn. Again, the type of safety shoe to be worn depends on the type of work that the worker is assigned to do.

An important factor about safety footwear that the plant safety official needs to keep in mind is that although OSHA requires employers to ensure that workers wear proper safety shoes on the job, OSHA does not mandate that the employer provide the safety shoes. This can be a point of contention between the worker and management. Safety shoes are not inexpensive. Some workers may resist purchasing expensive safety shoes. This is where the plant safety official must use the power of persuasion to ensure that management supports and workers comply with this important safety requirement.

Training on footwear protection is crucial. Kuhlman (1989) points out that those who suffered foot injuries had received no training on foot protection.

Safety shoes come in a variety of styles and materials, such as leather and rubber boots and oxfords (see Figure 3.22).

ELECTRICAL SAFETY (29 CFR 1910.301–.339)

Wastewater workers seem to have a healthy respect for electricity, which is well warranted. In 1985, for example, the Bureau of Labor Statistics reported that 338 deaths were the direct result of electrocutions at work. What makes these deaths more tragic is that, for the most part, they could have been avoided.

In a way, it seems ironic that most workers express a deep respect for electricity, yet seem to ignore or abuse electrical safe work practices. Perhaps the answer lies in the fact that electricity, as a source of power, has become accepted without much thought to the hazards encountered. That is, because it has become such a familiar part of our surroundings, it often is not treated with the level of respect and/or fear it deserves.

Wastewater treatment and collection workers are exposed to electrical equipment and hazards on a daily basis. For this reason, the wastewater

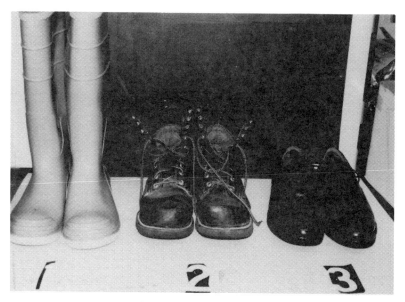

FIGURE 3.22. Safety footwear.

safety official must pay particular attention to this important safety topic. When seeking information and guidance on electrical hazards and their control, several sources of information are available. For example, OSHA has devoted Subpart S of its 1910 manual to rules governing electrical work. OSHA requires the employer to train all workers in safe work practices for working with electrical equipment.

OSHA's training rules distinguish between workers who do not work on or near exposed energized components and those who do. Even if the worker is not qualified to work on or around energized electrical equipment, he/she is required to know the specific safety practices which apply to their jobs.

The key to developing and maintaining a sound electrical safety program is to make worker awareness of electrical hazards an important part of the safety and health program. Worker *awareness* is brought about through training. Specifically, workers need to be aware of the primary hazards of electricity and know that the primary hazards of electricity are shock, burns, fires, explosions, and arc blast.

They must also have knowledge of the causes of electrical accidents and must be trained to understand that accidents and injuries in working with or around electricity are caused by a combination of factors. Factors such as unsafe equipment, workplaces, work practices, and equipment installation can all lead to electrical shock or worse.

Along with ensuring worker training on and awareness about electrical hazards, the plant safety official must make elimination or control of electrical hazards his/her goal. Electrical accidents can be prevented through protective methods to prevent electrical hazards. These protective methods include ensuring that electrical devices and circuitry are properly insulated, including insisting that electrical grade matting be installed in front of all high voltage switchgears. Electrical protective devices such as fuses, circuit breakers, guardrails, and ground connections should also be utilized. The plant safety official should check the integrity of electrical wiring throughout the facility. Often electrical shock can be prevented if hanging live electrical wires are discovered and properly removed. When the safety auditor finds hanging live electrical wires, he/she needs to confront the source of the occurrence. In other words, how did such an unsafe condition get left "hanging around," so to speak? When such conditions are found, it is instantly obvious that safe electrical work practices are being ignored.

Wastewater treatment plant electricians must also be trained in safe work practices. While it is usually true that electricians are highly trained and skilled technicians, it is also true that they can make mistakes, use poor judgment, or perform their work in a careless manner. Electrical safety must be stressed to *all workers.* Moreover, all workers must be reminded that the plant requires the use of a lockout/tagout procedure to de-energize electrical equipment and that it will be enforced. Additionally, workers must be reminded not to wear metal objects (watches, rings, etc.) when working with electricity.

In summary, an effective workplace electrical safety program can be instituted if the following steps are followed and enforced:

(1) Identify and label all workplace electrical hazards (Figure 3.23).

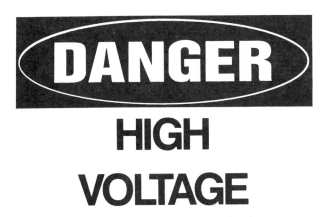

FIGURE 3.23. Electrical hazard warning sign.

(2) Train all employees on electrical safety.
(3) Use safe work practices. These include
- de-energizing electric equipment before inspecting or making repairs
- using electrical tools that are in good repair
- using sound judgment when working near energized lines
- using appropriate protective equipment

NOTE: In Chapter 7 of this text more will be said about electrical safe work practices.

FIRE SAFETY (29 CFR 1910.38 AND 1910.157)

Wastewater treatment plants are not immune to fire and its terrible consequences. Fortunately, the plant safety official is aided in his/her efforts in fire prevention and control by the authoritative and professional guidance that is available from the National Fire Protection Association (NFPA), the National Safety Council (NSC), Fire Code Agencies, Local Fire Authorities, and OSHA regulations.

Along with providing fire prevention guidance, OSHA regulates several aspects of fire prevention and emergency response. For example, emergency response, evacuation and fire prevention plans are required under OSHA's 1910.38. Additionally the requirement for fire extinguishers and worker training are addressed in 1910.157. More specifically, along with state and municipal authorities, OSHA has listed several specific fire safety requirements for general industry.

All of the just-mentioned advisory and regulatory authorities approach fire safety in much the same manner. For example, they all agree that fires in the workplace are usually started by electrical short circuits or malfunctions. Other leading causes of fire in the workplace are friction heat, welding and cutting of metals, improperly handled chemicals, improperly stored flammable/combustible materials, open flames, and cigarette smoking.

For fire to start, three conditions or ingredients are needed: *heat, fuel,* and *oxygen* (see Figure 3.24).

The fire triangle helps one understand fire prevention. For it is the objective of fire prevention and firefighting to separate any one of the fire ingredients from the other two. For example, to prevent fires, keep fuel (combustible materials) away from heat (as in airtight containers), thus isolating the fuel from the oxygen in the air.

To help gain a better perspective of the chemical reaction known as *fire,* it should be pointed out that the combustion reaction normally occurs in the

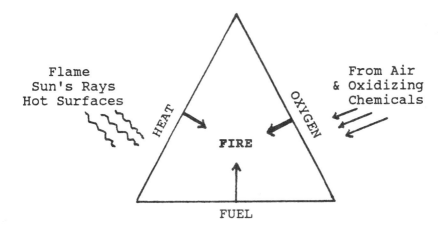

Paper-Wood-Fabrics
Gasoline-Alcohol-Solvents

FIGURE 3.24. Fire triangle.

gas phase; generally, the oxidizer is air. If a flammable gas is mixed with air, there is a minimum gas concentration below which ignition will not occur. That concentration is known as the *Lower Flammable Limit (LFL)*. When trying to think of LFL and its counterpart, *UFL (Upper Flammable Limit)*, it helps to use an example that most people are familiar with – the combustion process that occurs in the automobile engine. When an automobile engine has a gas/air mixture that is below the LFL, the engine will not start because the mixture is too lean. When the same engine has a gas/air mixture that is above the UFL, it will not start because the mixture is too rich (engine is flooded). However, when the gas/air mixture is between the LFL and UFL levels, the engine should start.

Fire experts advise that the best way to reduce the possibility of fire in the workplace is prevention. The plant safety official has an important responsibility in fire prevention. In the first place, he/she must ensure that workers are trained in fire prevention practices. Second, workplace housekeeping practices must be monitored and strongly enforced. If workers are allowed to let debris or flammable material accumulate, the risk of fire increases. The plant safety official and work center supervisors must ensure that all workers understand that fire prevention is everyone's job. Because of the deadly nature of fire, it is to all workers' benefit to know how to size up a fire and how to respond in a fire emergency.

Fire Prevention and Control

Fire prevention and control measures are those taken *before* fires start. Fire prevention and control is best accomplished by

- elimination of heat and ignition sources
- separation of incompatible materials
- adequate means of fire fighting (sprinklers, extinguishers, hoses, etc.)
- proper construction and choices of storage containers
- proper ventilation systems for venting and reducing vapor buildup
- in the event of fire emergency, unobstructed means of egress for workers. Additionally, adequate aisle and fire-lane clearance for firefighters and equipment must be maintained.

In the case of a fire emergency, all workers need to know what to do; they need a plan. The Fire Emergency Plan normally is the protocol to follow for fire emergency response and evacuation. Usually, the plant safety official is charged with developing fire prevention and emergency response plans that spell out everyone's role. Make your fire plan as easy as possible. Along with an easily understood fire plan the workers need to know what actions they are expected to take in the event of a fire. Figure 3.25 shows an example of a simple decision-tree format that is used by wastewater treatment and collection facilities for fire emergencies.

In addition to some type of fire emergency response action plan like the one shown in Figure 3.25, each plant needs to have a well-thought-out fire emergency evacuation plan.

Fire Protection

OSHA requires employers to provide portable fire extinguishers that are mounted, located, and identified so they are readily accessible to workers without subjecting the worker to possible injury. In addition, OSHA requires that each workplace institute a portable fire extinguisher maintenance plan. Fire extinguisher maintenance service must take place at least once a year and a written record must be kept to show the maintenance or recharge date. *NOTE:* When the plant safety official provides portable fire extinguishers for worker use in the plant, the worker must be provided with an education program to learn the general principles of fire extinguisher use and the hazards involved in fire fighting.

Wastewater workers must be trained to know what type of fire extinguishers are available for the different classes of fire and where they are located. The ABC type fire extinguisher is probably best suited for the

FIRE EMERGENCY

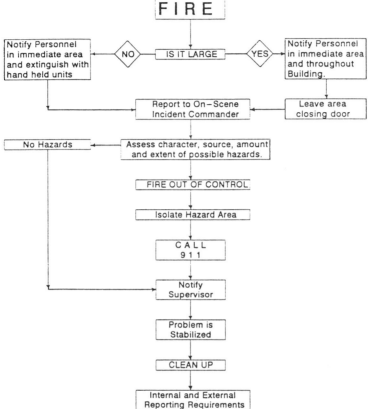

FIGURE 3.25. Fire emergency decision tree.

wastewater industry because it can be used on Class A, B, and C fires. The only exception is electrical substations or switchgear rooms. In areas like these, place Class C (Carbon Dioxide–CO_2) extinguishers only. Combination Class A, B, and C fire extinguishers will extinguish most electrical fires, but the sticky chemical residue left behind can damage delicate electrical/electronic components; thus, CO_2 type extinguishers are more suitable for extinguishing electrical fires.

Each worker must know how to use the fire extinguisher. Most importantly, the workers must know when it is *not safe* to use fire extinguishers; that is, when the fire is beyond being extinguishable with a portable fire extinguisher.

Emergency telephone numbers should be strategically placed throughout the workplace and workers need to know where they are posted. Moreover, workers should be trained on the information that they need to pass on to the 911 operator in case of fire.

In addition to basic fire prevention and emergency response training, workers must be trained on flammable and combustible liquids. OSHA Standard 29 CFR 1910.106 addresses this area.

Wastewater treatment and collection operations use all types of flammable and combustible liquids. These dangerous materials must be clearly labeled and stored safely when not in use. Additionally, the safe handling of flammable and combustible liquids is a topic that needs to be fully addressed by the plant's safety official and workplace supervisors. Worker awareness of the potential hazards that flammable and combustible liquids pose must be stressed. Workers need to know that flammable and combustible liquid fires burn extremely hot and can produce copious amounts of dense black smoke. In addition, explosion hazards exist under certain conditions in enclosed, poorly ventilated spaces where vapors can accumulate. A flame or spark can cause vapors to ignite creating a flash fire with the terrible force of an explosion.

One of the keys to reducing the potential spread of flammable and combustible fires is to provide adequate containment. All storage tanks should be surrounded by storage dikes or containment systems, for example. Correctly built and installed dikes will contain spilled liquid. Spilled flammable and combustible liquids that are contained are easier to manage than those that have free run of the workplace. Additionally, properly installed dikes can prevent environmental contamination of soil and groundwater.

In summary, the plant's workers, supervisors, and safety official must be prepared for fire and its consequences. It is important for the plant to maintain a fire prevention strategy that will ensure that work areas are clean and clutterfree. Workers must know how to handle and properly store chemicals, what they are expected to do in case of a fire emergency, and how and whom to call when fire occurs. Workers must be thoroughly trained in fire prevention and emergency response procedures.

LABORATORY SAFETY—OSHA, CERCLA, AND SARA

Wastewater treatment plant laboratories function to perform process-control required tests for permit compliance. The size of the laboratory generally depends on the size of the plant. For example, some plants are satellite treatment works within a larger sanitation district. In such cases, each individual plant usually has a laboratory designed specifically to handle the work required to meet compliance and perform testing criteria for the plant. On the larger scale, a sanitation district's main laboratory usually conducts testing for the entire organization; it augments and refines site testing.

No matter the size of your wastewater treatment laboratory, safety plays a key role in maintaining worker well-being and compliance with applicable Health, Safety, and Environmental Standards. Wastewater treatment laboratories are expected to comply with OSHA, under 29 CFR 1910.145, which mandates a Chemical Hygiene/Chemical Hazard Plan (CHP). Additionally, wastewater treatment laboratories must comply with the requirements of the Comprehensive Environmental Response, Compensation, and Liability Act (CERCLA), with Superfund Amendments and Reauthorization Act (SARA) rules and regulations, as well as with rules and procedures set by the State Water Control Board or other local environmental agencies.

The wastewater treatment plant safety official should ensure that the laboratory has a safety program. The goal of the laboratory safety program is multifaceted: It should protect laboratory workers, others who may be exposed to hazards from the laboratory, and the environment.

The hazards associated with laboratories are not unlike those in other work areas. Common injuries include cuts, burns, slips, and falls. However, because of constant exposure to potentially dangerous substances, all personnel that work in the laboratory must be instructed on proper laboratory methods and safety precautions.

Laboratory personnel are required to perform analyses on a variety of wastewater and biosolids (sludge) samples. While it is true that the risk of infection from wastewater samples is not as high in the laboratory environment as in the collection systems, where direct exposure to raw wastewater is high (in sewers, pumping stations, and interceptor pipelines), infectious agents commonly found in laboratory samples are nonetheless a biological hazard.

Very specific safety procedures should be followed in the wastewater treatment plant laboratory. Chapter 7 lists a specific Safe Work Practice Procedure for laboratory personnel. In addition to a rigid safety protocol, laboratory workers must learn to refer to Material Safety Data Sheets (MSDS) whenever they need information on any chemical used in the laboratory. Because of the inherent hazards of wastewater treatment laboratory work, in addition to the safe work practice in Chapter 7, safe laboratory procedures are provided in the following.

Laboratory Safety Procedures

INFECTIOUS MATERIALS

Raw sewage and biosolids contain millions of bacteria, some of which are infectious and cause disease such as tetanus, dysentery, poliomyelitis, or hepatitis. *NOTE:* Whenever hepatitis is mentioned today, bloodborne pathogens like HIV and Hepatitis B come to mind. All workers should protect against these deadly diseases. However, in the wastewater industry, the Center for Disease Control (CDC) has found no definitive proof that HIV and Hepatitis B are transmitted by the wastewater stream. These two diseases have been detected and sustained in the wastewater stream *only* under strictly controlled laboratory conditions. Always wash your hands with soap and water, particularly before handling food or drink. Use rubber gloves when handling sewage samples; cuts or abrasions in the skin can easily allow infection to enter the body.

CORROSIVE CHEMICALS

Sulfuric, hydrochloric, nitric, glacial acetic, perchloric acids, and chromic acid cleaning solutions quickly destroy human tissue, clothing, and wood on contact. They are also extremely corrosive to metals and concrete. Use glassware or polyethylene containers. Always add acid to water, not vice versa. Avoid contact with metals. Pour and pipette acid carefully. Never pipette by mouth; use a pipette bulb. If spills occur, follow these safety procedures:

- Dilute the affected area with large portions of water. Clean up the diluted material.
- For spills on bench tops, wear gloves and dilute and squeegee into the sink.
- Human contact: immediately wash off with large quantities of water.
- If spills of concentrated acid splash on your face, flood with large amounts of cold water. Notify your supervisor.
- If acid gets into your eyes, use the emergency eye wash stations and flush for at least 15 minutes; seek immediate medical attention.

Bases such as sodium hydroxide, potassium hydroxide, and ammonium hydroxide quickly digest skin, clothing, and leather on contact. Always handle bases with extreme care. Use glassware or polyethylene containers. Ammonium hydroxide is extremely irritating to the eyes and respiratory system. When pouring ammonium hydroxide, use a fume hood with the ventilation on. In case of an accident involving bases, wash the area with large quantities of water until the slippery feeling is eliminated. When handling chlorine gas solution, also use a fume hood and avoid vapor escape. Secure the fume hood cover to prevent vapor escape. Ferric salts and ferric chloride are very corrosive to metals. Avoid bodily contact with them and wash off any residue immediately.

TOXIC MATERIALS

Apply the following safety procedures when working with toxic materials:

- Read all toxicity warnings and antidote information on labels of laboratory chemicals. Learn about the materials that you are handling by reading the MSDS.
- Avoid ingesting or inhaling any chemicals, especially cyanide, chromium, cadmium, and other heavy metal compounds.
- Use a vented hood when handling ammonium hydroxide, nitric acid, bromine, ammonia, aniline dyes, formaldehyde, chloroform, and carbon disulfide. Moreover, remember that carbon tetrachloride is absorbed into the skin on contact and is cumulative.

EXPLOSIVE OR FLAMMABLE MATERIALS

Bottled gases such as acetylene and hydrogen, and liquids such as carbon disulfide, benzene, ethyl ether, petroleum ether, and acetone, all require careful handling. Always store them in accordance with proper fire prevention procedures, as listed on the MSDS.

Follow these procedures when handling explosive or flammable materials in the laboratory:

- Use a vented hood.
- Never use explosive or flammable materials near open flame or expose to electrical heating elements.
- Do not distill to dryness, or explosive mixtures may result.

GLASSWARE

Broken glassware in the laboratory can lead not only to workers getting cut but also to unnecessary chemical exposure. Follow these simple safety procedures when handling glassware in the laboratory.

- Immediately dispose of all broken or cracked glassware in the designated disposal container. Chipped glassware may be used if it can be fire polished to eliminate sharp edges.
- Use caution when making rubber-to-glass connections. Support lengths of glass tubing while they are being inserted into rubber. Fire polish ends of glass tubing until smooth. Use a lubricant such as water—*never use grease or oil*. Use gloves when making such connections and hold tubing as close as possible to the end being inserted to prevent bending or breaking. Never try to force rubber

tubing or stoppers from glassware; cut rubber as necessary to remove.

Summary of Safe Procedures in the Laboratory

Usually, personnel assigned to wastewater treatment facility laboratories are chemists, biologists, environmental scientists, or ecologists. These personnel are usually well trained in safe laboratory procedures. The key word here is "usually." Actually, smaller (and some larger) wastewater laboratories may use plant operators to perform basic laboratory testing procedures. Whether your laboratory is staffed by skilled professionals or technicians, safety is still a major concern in all wastewater laboratories.

Safety in the laboratory begins with training. New employee indoctrination or orientation is critical. This topic will be addressed in detail later. For now it should be noted that without indoctrinating the new employee to the hazards that are present within the plant and the plant laboratory, the employee is not properly prepared to begin work. In addition, OSHA, in its Hazard Communication Standard, is very specific about not putting a new employee to work with or around hazardous materials without first providing proper training.

Hazard communication training is only the first step. Laboratory workers must be informed of laboratory rules and safety procedures. Only authorized, properly trained persons should be allowed in the wastewater treatment plant laboratory. Personal protective equipment (PPE) must be made available and required to be used in the laboratory. All chemicals must be labeled in accordance with the Hazard Communication Standard and the Chemical Hygiene Plan. Accurate recordkeeping procedures must be used to ensure that the date that chemicals are opened and safely disposed of is recorded. The laboratory must have a broken glassware procedure designed to ensure the safe disposal of broken or cracked glassware.

Good housekeeping practices must be enforced at all times. All work areas and other equipment must be kept clean. Clutter must be avoided, and unused items stored in their proper places. The laboratory should never be used as the plant social center. Moreover, food should never be eaten or stored in the laboratory.

The laboratory should be fully equipped with emergency eye wash, shower, and spill neutralization kits to wash out chemical spills. The plant safety official should ensure not only that eye wash and emergency showers are present within easy access of laboratory workers, he/she should also ensure that routine preventive maintenance is performed on this vital safety equipment.

Laboratories generally contain several types of portable electrical measuring devices. These electrical meters and other instruments should not be used in close proximity to water and sinks. Fume hoods should be used for opening samples and conducting tests; they should not be used for storage.

The plant safety official must pay close attention to laboratory activities. Laboratory workers who have been properly trained in laboratory safety are a definite plus. A laboratory that is properly equipped with safety equipment, spill neutralization and first aid kits is a workplace where activities can be conducted in safe fashion. It is important to remember that the Hazard Communication Standard not only requires labeling of chemical hazards but also labeling of hazard areas. The first item that any inspector looks for in a laboratory is labeling. The inspector looks to see if fire extinguishers, eye wash/emergency showers, chemical spills kits, acid/corrosive storage areas, poison control lockers and first aid stations are present, properly maintained, and labeled.

As part of the laboratory's Chemical Hygiene Plan (CHP), emergency response is critical. Written procedures are required for neutralizing, controlling, and mitigating chemical spills. Along with written procedures for emergency response, laboratory workers must be thoroughly trained on the correct procedures.

These procedures must be monitored. Observations of individual safety practices, operability of safety equipment, and compliance with safety rules should be part of this monitoring process. Any malfunctioning machine or equipment should be reported and repaired. Essential safety equipment, such as sterilizers and eye wash fountains, should be tested periodically and a record kept of their last inspection (to be discussed in greater detail later). PPE for use in an emergency should also be checked periodically.

Spills and Releases in Laboratory

Accidental spill and release of hazardous substances is a common occurrence in the wastewater laboratory. Therefore, procedures for minimizing exposure of personnel and contamination of property are required. Such procedures may range from common cleanup materials (sponges, rags, mops, etc.) to having an emergency response team, complete with all required emergency response equipment (PPE, safety equipment, and chemical spill kits). The emergency response spill/release action plan should be contained in the laboratory's Chemical Hygiene Plan. It is important that all laboratory workers are trained on the Plan.

SLIPS, TRIPS, FALLS, AND SAFE LIFTING PRACTICE

Slips, trips, and falls are common mishaps in wastewater treatment facilities. Many workplace activities involve carrying loads and other material handling activities, which are major contributors to slips, trips, falls and back injuries. Additionally, the use of ladders or scaffolds can put workers at risk for fall-related injuries.

When chemicals are introduced into the work activity, the occurrence of slips, trips, and falls seems to escalate. As a case in point, consider chemical polymers. Polymers are used in wastewater treatment for varying purposes. One such use is for biosolid (sludge) conditioning. When polymer becomes wet, its slippery nature produces a severe walking surface hazard.

Polymer is only one of several chemicals used in wastewater treatment facilities that can create slippery walkways and stairs. Therefore, all spills must be cleaned up immediately. Workers must be trained to react whenever they spill chemicals or observe a spill. Even minor spills can be hazardous.

In practicing good housekeeping in the workplace, the maintenance of properly cleaned floors is step one. Providing well-lighted work areas is another step. Additionally, clutter in aisles or stairs is a major contributing factor to trips and falls. Hazards due to loose footing on stairs, steps, and floors must be eliminated.

Since the plant safety official plays a major role in preventing slips, trips, and falls in the workplace, his/her routine plant audit should focus on this vital area. Moreover, plant safety meetings should always stress the importance of keeping the work area clean.

Ranking right up there in frequency of occurrence to finger and hand injuries are injuries to the back. Back injuries present a constant dilemma to the plant safety official. How are they to be prevented? What steps are necessary to help prevention?

Back injury prevention begins with the safety plant official gaining an insight into mishaps caused by material handling. Such insight is gained through consideration of several factors. For example, it should be determined whether or not it is practical to provide workers with material handling aids that will make lifting safer. Lifting aids include insisting that workers lift the properly sized containers, use lifting trucks, and use special lifting tools such as hooks. Another aid to safe lifting is whether or not it is practical to provide conveyers or other mechanical devices for moving packages. As a last resort, the safety official should decide if personal protective equipment such as gloves might help prevent injuries.

After making an assessment of the task and the workplace, reviewing reports of back injuries that have occurred at the plant site in the past, and determining whether or not mechanical devices or personal protective equipment can help reduce such back injuries, the safety official should decide which type of worker back injury prevention training program to institute. Experience has shown that training can be an effective management aid in helping reduce back injuries.

The wastewater treatment plant back injury prevention training program should stress the following key elements in avoiding on-the-job back injuries from lifting tasks:

- Bend your knees—not your back.
- Exercise to strengthen muscles and improve flexibility.
- Plan each lift prior to starting.
- Let the legs, not the back, power the lift.
- Lift smoothly—never twist while lifting.
- Don't overdo.

The back injury prevention program is ongoing. Back injury prevention and other types of safety training must be presented on a continuous basis. Moreover, followup training is the key to maintaining what was learned in the initial training sessions.

EXCAVATION SAFETY (29 CFR 1926.650–.652)

According to Stanevich and Middleton (1988), about 50% of trenching and excavation specialty companies (e.g., utilities) experience cave-ins. Wastewater collection systems are an integral part of wastewater treatment, which is a utility that commonly performs trenching and excavation work because maintaining a leak-free, infiltration-free interceptor piping system is important to providing a constant wastewater stream to the treatment facility. However, pipes fail. When they fail, they must be repaired or replaced. On occasion, before repairing or replacing the lines, they must be excavated.

Repair work to above-ground interceptor lines is relatively easy compared to making repairs to underground lines. When repairs are to be made to underground lines, either trenching or excavation is required. A trench is a narrow excavation that is less than 15′ wide and is deeper than its width. An excavation, on the other hand, is a cavity or depression that is cut or dug into the earth's surface.

Whether trenching or excavating, either operation is inherently dangerous. As a matter of fact, working around and in excavations is one of the most dangerous jobs in the wastewater treatment and construction industries. In the construction industry alone, it is estimated, for example, that cave-ins claim about 100 lives every year. Thus, excavating interceptor lines is a very real hazard to wastewater workers. Past incident investigations of trenching and excavation mishaps have shown that little heed was paid to the hazards involved with excavating and trenching.

If your facility makes routine repairs or replacement to underground interceptor lines, you must ensure that the OSHA regulations pertaining to trenching and shoring are followed. These OSHA regulations can be found in 1926.650–.652 (the Construction Standard).

An effective trenching and shoring safety program begins with *knowing the hazards.* Workers must know what they face during these operations. Additionally, they must know how to protect themselves from injury through proper use of safe work practices – in trenching and excavation work there is no room for error. When a trench or excavation fails, injuries and fatalities occur fast – often in seconds.

There is no room for poor judgment in performing excavation and trenching operations. Before beginning to trench or excavate, certain conditions must be checked. For example, prior to beginning an excavation the location of utility installations, such as telephone, fuel, electric, water lines, or any other underground installations that may be expected to be encountered during excavation work must be determined.

Other conditions must also be checked before the excavation is attempted. For example, soil, weather, and climate conditions are important factors to be checked. These factors will determine the amount and degree of sloping that should be used. Moreover, the actual strength of trenching support members (bracing and shoring) is based on soil type and weather conditions.

Failure to properly support walls for a trench or excavation may cause disaster. Often trenching and excavation jobs are driven by cost- and time-saving requirements. At other times, the supervisor in charge of the operation determines that since the trench or excavation will only be open for a short period, proper shoring is not needed. What this supervisor is really doing is taking a chance – a shortcut. The result may be a shortcut to disaster.

In auditing plant excavation programs, it is not unusual to find a weak link. Usually, this weak link can be attributed to a lack of supervisor and worker knowledge. Thus, the auditor often finds that workers are not made aware of the hazards involved and the precautions necessary to minimize the hazard.

A trenching and excavation training program should provide the information necessary to ensure that supervisors and workers know the hazards. They should know that the major hazard is cave-ins, which can crush or suffocate them. They should know that trenches and excavations can contain poisonous gases – there is the very real possibility of uncovering a pocket of combustible vapors or gases. For example, when digging around interceptor lines, it is not unusual to run into a pocket of methane, leading to the danger of fire and/or explosion.

Supervisors and workers also need to know that OSHA requires excavations to be protected from cave-ins by an adequate protective system designed to resist, without failure, all loads that are intended or could reasonably be expected to be applied or transmitted to the system.

Trenches and excavations can be full of additional obstacles. For example, carelessly placed tools and equipment or excavated material can cause injuries due to slips, trips, or falls. After explaining the general hazards that are present in any trenching or excavation job, the person performing the training must inform the workers about the causes of cave-ins – they need to know what to look for. Workers need to understand that cave-ins occur when an unsupported wall is weakened or undermined by too much weight, or pressure, or an unstable bottom.

One of the danger signs to look for in trenching or excavation work is surface cracking. Cracks usually occur near the edge of the trench or excavation. Overhangs and bulges are other signs of danger. An overhang at the top – or a bulge in a wall – can cause soil to slide into the trench or excavation. Whenever cracks or overhangs are discovered, work should be stopped and the problem reported immediately.

As mentioned earlier, weather and climate can have a serious impact on trenching and excavation activities. Rain, melting snow, ground water, storm drains, nearby streams and/or damaged water lines can loosen soil and increase pressure on walls. On the opposite extreme, extremely dry weather can also be dangerous because it tends to loosen soil. Frozen ground presents another problem. When the frozen ground thaws, the walls of a trench or excavation can be weakened. Especially when the excavation is a long-term job, there may be a need for extra weather and climate protection. Sides and faces of the dig should be covered with tarps to reduce danger.

Supervisors and workers also need to be trained in soil-type recognition procedures. Soils with high silt or sand content are very unstable, unless properly shored or sloped. Wet or back-filled soil is also unstable and needs wall support. Even hard rock can present a problem unless it is properly supported. For example, hard rock that cracks or splits through a fault can break away and fall into the excavation.

One area of danger that is often overlooked in trenching and shoring operations is the presence of vibration. Vibrations can loosen soil and cause walls to collapse unless proper shoring or sloping is used. There are several sources of vibration at the work site: vehicles, moving machinery, blasting operations, and machines that might be used nearby such as punch presses and forging hammers.

Excavated material can also pose a hazard to the excavators. Excavated material (or spoil) should always be stored at least 2' from the edge of a trench or excavation. Never let excavated material accumulate near wall sides. Additionally, moving excavated soil can also pose a hazard to excavators. Heavy equipment operating near the trench or excavation can exert tremendous pressure on walls.

In order to protect excavators against accidents, proper techniques and equipment must be used in the trenching and shoring operation. Trench shoring material should consist of sheeting, bracing, and jacks. Never use shoring materials that have not been certified for use by a licensed professional engineer. After installation of the correct shoring materials, these materials should be inspected daily before anyone enters the trench or excavation.

When the decision is made to use ground sloping techniques to prevent cave-ins, the sides of a trench or excavation must be sloped correctly so that soil will not slide. Determining the *angle of repose* is critical in shaping proper slope. The angle of repose is the steepest angle at which trench or excavation walls will lie without sliding. The more unstable the soil, the flatter the angle should be. For example, the angle of repose for solid rock should be set at 90°. For average soil an angle of 45° is recommended. For loose sand the proper angle of repose should be set at about 25°.

Several other safety considerations are necessary whenever a trenching or excavation project is undertaken. Workers must understand that it is important to provide site protection. Site protection not only protects workers from rocks or other objects kicked or thrown into the trench but also protects pedestrians who might inadvertently fall into an open trench or excavation. Safety measures such as fences, barricades, covers for manholes, flags, security guards and warning signs may be necessary. Lighting is sometimes necessary for maintaining safety at night.

Further, a stairway, ladder, ramp or other safe means of egress must be located in trench excavations that are 4 feet or more in depth so as to require no more than 25 feet of lateral travel for workers.

As with any other dangerous operation, the plant safety official should ensure that a contingency plan for emergency response is used. This contingency plan must be made clear and understandable to all trenching and excavation personnel. Emergency procedures are worthless unless they are common knowledge. Someone should always be outside the trench or excavation to help, if necessary. Emergency telephone and numbers should be readily available. As a final precaution, the trench or excavation should be backfilled as soon as possible when the work is completed.

As with many other work activities, OSHA requires that personnel involved with excavation activities be trained and that this training be documented. Moreover, OSHA requires that the person in charge of the excavation be a qualified or competent person. When qualifying a qualified/competent person for excavation/trenching operations, it is wise to put together an excavation crew with a "potential" qualified/competent person as the person in charge and have the crew dig a 12 foot deep hole in a practice area. Experience gained through actually performing the work can never be replicated by listening to a classroom lecture on the topic.

While observing the dig and shoring procedures, note whether or not the potential qualified/competent person is conducting the dig as required. If the dig is done correctly, certify the person in charge as a Qualified Person for Excavation/Trenching Operations (see Figure 3.26). The other excavation crew members should be given a Certificate of Training verifying that they have been trained.

HAZARDOUS MATERIALS EMERGENCY RESPONSE (29 CFR 1910.120/40 CFR 311)

CoVan (1995) defines emergency response "as a limited response to abnormal conditions expected to result in unacceptable risk requiring rapid corrective action to prevent harm to personnel, property, or systems function" (p. 54). OSHA and EPA have mandated that facilities that handle or use hazardous materials (e.g., chlorine, sulfur dioxide, sodium hydroxide, methane, etc.) develop a Site Emergency Response Plan and worker training in order to provide workers correct guidance on what to do in case of medical, fire, and/or chemical discharge emergencies. The OSHA/EPA joint standards on Hazardous Waste Operations and Emergency Response is commonly called *HAZWOPER*.

The HAZWOPER Standard impacts wastewater treatment plants. When

FIGURE 3.26. Certification for qualified person for excavating/trenching operations (*Source:* Hampton Roads Sanitation District; used by permission).

wastewater treatment plant management personnel are informed about this, they are often surprised. To judge whether or not HAZWOPER impacts them, plant managers should perform a site survey. Such a survey is designed to account for and list all hazardous materials. For example, if a wastewater treatment plant uses more than 10 pounds of chlorine in its process, it must be prepared to manage or mitigate an accidental release of chlorine to comply with the HAZWOPER Standard.

The Site Emergency Response Plan for wastewater treatment plants should include medical emergency instructions, fire emergency plans, and chemical release emergency plans. In addition, it should include an emergency evacuation plan, a chemical/safety equipment location diagram, and hazardous materials system line diagrams. Use of DOT's Emergency Response Guidebook as the primary reference manual is highly recommended.

The plant's Emergency Response Plan should be user friendly. Plans written by technical personnel are usually slanted toward a technical point of view to be understood and utilized by technical persons. This can defeat the intended purpose: quick, correct, and safe mitigation procedures. Moreover, making emergency response highly technical or complicated defeats the purpose of making the plan user friendly. Additionally, workers must be fully aware of the plan's requirements and their own individual responsibilities.

The site emergency response plan should be written around two main objectives: (1) To minimize the short-term or immediate hazards to the public, the responders and the environment; and (2) to ensure the recovery and long-term use of the affected plant site. Accomplishing the first objective may necessitate taking short-term actions that will delay or prolong accomplishment of the second objective; however, these objectives must be accomplished in sequence.

Responding to a release of chlorine, for example, requires extensive training in chlorine repair kit use, chemical protective equipment and respiratory protection. In addition, the plant emergency response team must coordinate its response efforts with other local emergency response teams. The plant must have a designated emergency response coordinator who is trained to coordinate the plant's response to chemical release.

The HAZWOPER Standard requires emergency responders to be trained and certified at one of five levels.

- Level 1: *First Responders at the Awareness Level* are workers who are likely to discover a hazardous materials release and have been trained to *sound the alarm*.
- Level 2: *First Responders at the Operations Level* are trained to

contain a release from a safe distance and keep it from spreading (*contain it*) without trying to stop it.

- Level 3: *Hazardous Materials Technicians respond aggressively* to stop the release of hazardous materials. These trained personnel are the ones who are qualified to stop a leak or close a valve.
- Level 4: *Hazardous Materials Specialists have detailed knowledge* of specific chemicals and provide on-site cooperation with government officials.
- Level 5: *On-scene incident commanders take control* of the incident scene during emergency response (this position is usually filled by the senior Fire Department person present at the scene).

NOTE: All emergency responders must either receive refresher training or demonstrate that they are able to perform their duties at least once a year.

If plant management decides to take no emergency response actions itself, but instead decides to call 911 or local emergency response personnel, then the level of training required for plant workers is Level 1: *First Responder at the Awareness Level* (personnel who discover the spill/fire, sound the alarm, and take no mitigation action themselves).

Training for Level 1 personnel must cover the following topics:

- knowing what hazardous materials are, the risk they pose and the possible outcomes of a release
- recognizing the presence of hazardous materials
- identifying released hazardous materials
- recognizing the need for additional resources and notifying the proper authorities
- functioning in the role of the Level 1 responder in the employer's Emergency Response Plan and using the Emergency Response Guidebook published by DOT

Like all other types of safety training, emergency response training must be documented. Moreover, when training emergency responders for qualification as Level 1 thru Level 3 responders, it is wise to administer a written test to ensure that the workers can demonstrate a working knowledge of the subject matter. After the entire qualification process is complete, that is, after Level 1 thru Level 3 classroom training and twenty-four hours of field training, certificates of training should be awarded to the qualified responders. Additionally, a copy of the training certification must be inserted into each worker's training record.

You can be absolutely assured of one thing: If your plant has a major chemical incident, some time during or after the incident the OSHA/EPA

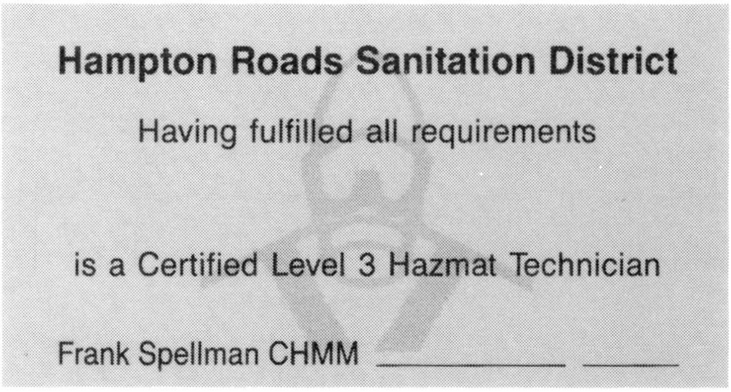

Hampton Roads Sanitation District

Having fulfilled all requirements

is a Certified Level 3 Hazmat Technician

Frank Spellman CHMM ____________ ______

FIGURE 3.27. HAZMAT responder certification card (*Source:* Hampton Roads Sanitation District; used by permission).

investigator will ask to see (among several other items) your training records. When the OSHA investigator asks the emergency responder what his/her level of HAZMAT training is and the responder can reach into his/her pocket and pull out the training certificate shown in Figure 3.27, you will feel a sense of relief that you might not have felt before. Those who have experienced this feeling of relief fully understand the benefits of their training and certification programs.

The Emergency Response Plan

The wastewater treatment plant's Emergency Response Plan states the organization's policies and procedures for dealing with emergencies. All workers should be familiar with the plan. Emergency Response Plans must be preapproved by local authorities when outside help, such as the local HAZMAT Team or fire department would normally be called to the scene to assist. The plan must list emergency phone numbers, which should also be posted near all plant phones.

If your plant decides to take an aggressive approach to handling chemical releases, the responders must be trained to Level 3. A few of the Level 3 training requirements were mentioned earlier. The training requirements for Level 3 HAZMAT Responders are quite extensive. At least 40 hours of hazardous materials response training are required on such topics as the law, toxicology, first aid, personal protective equipment, air monitoring, decontamination, site safety plans, sampling/chain of custody, and many more. At least 24 hours of this training must include hands-on sessions with chemical containment, spill cleanup procedures, and personnel decontamination procedures. A written certification examination is required.

The actual emergency response scenarios for a wastewater treatment plant are more effective and user friendly if they are limited to three contingencies: fire emergencies, medical emergencies, and chemical discharges. Figures 3.28 and 3.29 show two examples of simplified response scenarios.

From Figures 3.28 and 3.29 it should be obvious that these simplified procedures have several advantages. First, each procedure is contained on one page, allowing the responder to focus on those key action items that need to be initiated quickly. Second, since each procedure is fully contained on one page, training is less complicated. If a training document is too lengthy or bulky, it may not be used by workers. Third, the procedures can

MEDICAL RESPONSE GUIDE

Immediate Procedures

- Remain calm.
- Initiate lifesaving measures if required.
- Call for **Emergency Response:**

- Do not move injured person unless there is danger of further harm.
- Keep injured person warm.

MEDICAL EMERGENCY   911

EMERGENCY RESPONSE PROCEDURES

Clothing on Fire

- Roll person around on floor to smother flame,

  or

  drench with water if safety shower is **immediately available.**
- Obtain medical attention, if necessary.
- Report incident to supervisor.

Chemical Spill on Body

- Flood exposed area with running water from faucet or safety shower for at least 5 minutes.
- Remove contaminated clothing at once.
- Make sure chemical has not accumulated in shoes.
- Obtain medical attention, if necessary.
- Report incident to supervisor.

Biological Spill on Body (Lab Workers)

- Remove contaminated clothing.
- Vigorously wash exposed area with soap and water for 1 minute.
- Obtain medical attention, if necessary.
- Report incident to supervisor.

Hazardous Material Splashed in Eye

- Immediately rinse eyeball and inner surface of eyelid with water continuously for 15 minutes.
- Forcibly hold eye open to ensure effective wash behind eyelids.
- **Obtain medical attention.**
- Report incident to supervisor.

Minor Cuts and Puncture Wounds

- Vigorously wash injury with soap and water for several minutes.
- **Obtain medical attention.**
- Report incident to supervisor.

FIGURE 3.28. Medical emergency response plan (*Source:* Hampton Roads Sanitation District; used by permission).

CHEMICAL DISCHARGE

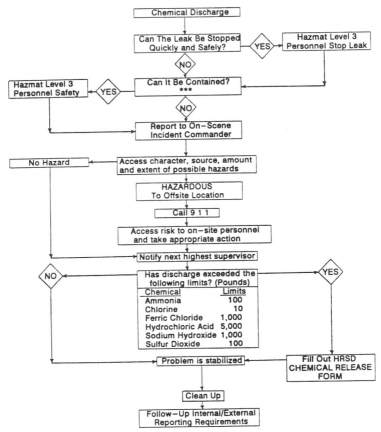

* Contain is defined as the closure of all doors, windows, louvers, and valves to the extent possible and practicable.

FIGURE 3.29. Emergency response: chemical release/spill (*Source:* Hampton Roads Sanitation District; used by permission).

be laminated, bound with a ring holder and placed at critical locations throughout the plant. In an emergency situation it is important to have critical information available *instantly.*

Another advantage of using the decision-tree system depicted in Figure 3.29 and in Figure 3.25 (Fire) is that the proper sequence of decision-making is provided. If each branch of the decision tree is followed from start to finish, chances are excellent that the correct response actions will be taken.

Another important emergency response item for chemical spills/releases is the *site map*. The site map should point out in simple fashion exactly where on the plant site the hazardous materials (gasoline, fuel oils, propane, chemicals, etc.) are stored. A properly crafted site map can save emergency response units, local fire or HAZMAT team, valuable time.

In addition, having a simplified chemical system line diagram readily available is important. Here again, the importance of simplification cannot be overemphasized. Chemical systems and their associated piping and valve networks can be very complicated to follow. Emergency response chemical system line diagrams should be as simple as possible. Figure 3.30 shows a simplified line diagram for a digester methane gas system at one of HRSD's treatment plants.

The key item on the line diagram is the location of the "critical valve." That is, the valve that will shut down or isolate the system in an emergency. *NOTE:* When planning for proper emergency response operations, it is important to be able to estimate how much material may be set free or lost if there is a leak in the piping. For example, if a wastewater treatment plant has a large digester, there is an excellent possibility that large quantities of methane may be present within the digester. The emergency response planner must remember that there is also the possibility for large amounts of this dangerous substance to be inside piping associated with the digester. Both the leak rate and the total amount leaked are critical in forecasting the consequences of the leak. Thus, piping systems containing hazardous materials, (e.g., methane) should be clearly labeled along with all *critical valves*.

The critical valve must be marked and clearly visible for easy identification in both daylight and in darkness. This is an important point, especially since the emergency response action may be provided by a fireman or HAZMAT responder who is not familiar with the plant. One other point, the chemical system line diagram should be placed in close proximity to the critical value—but not too close. The line diagram is of little value if it is consumed or damaged during a fire or explosion.

PROCESS SAFETY MANAGEMENT (PSM) (29 CFR 1910.119)

The Process Safety Management Standard (PSM) was promulgated by OSHA and has been in effect since February 1992. The regulation serves to protect workers from accidents due to the release of *highly hazardous chemicals* (i.e., those chemicals listed under the regulation).

OSHA Booklet 3132, page 4, states that both municipal and industrial wastewater treatment plants are covered under the Process Safety Management (PSM) Standard. Moreover, several chemicals used and generated at

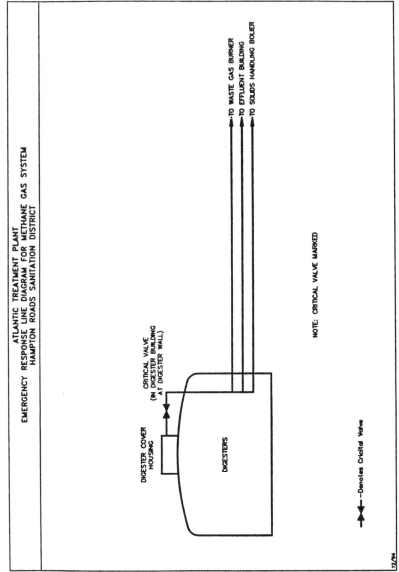

FIGURE 3.30. System line diagram (*Source*: Hampton Roads Sanitation District; used by permission).

104

wastewater treatment plants are covered by the PSM Standard (see Figure 3.31).

If the process in the plant contains any of the chemicals listed in Figure 3.31 at quantities at the threshold or greater, a Process Safety Management Program must be implemented. The threshold quantity is defined as the total quantity of material in the process that can potentially be released during an abnormal process situation (CAE Consultants, 1993). The purpose of OSHA's PSM requirement is to prevent unwanted releases of hazardous chemicals from covered processes.

Process Safety Management for wastewater treatment plants is designed to help the plant function safely. Moreover, PSM is designed to promote better quality through elimination of process fluctuations and process shutdowns. This is accomplished by eliminating unsafe changes from normal operating practice. Additionally, PSM is designed to identify operations that are not being done properly or are not operating properly. Through PSM a determination is made on how to perform operations safely.

PSM requires wastewater treatment plants to develop and maintain written safety information about hazardous processes. This information should

CHEMICAL	THRESHOLD QUANTITY
Chlorine	1500 lbs
Sulfur Dioxide	1500 lbs
Biogas	10,000 lbs
Sludge	10,000 lbs
Ozone	100 lbs
Flammables (Polymers,etc.)	10,000 lbs
Hydrogen Sulfide	1500 lbs
Ammonia Solution	10,000 lbs
Flammable liquids or gases such as propane, methane, oil, etc.	10,000 lbs
Hydrogen peroxide	@52% 7500 lbs
Chlorine Dioxide	1000 lbs
Hydrochloric Acid	5000 lbs
Oxygen	10,000 lbs

FIGURE 3.31. Chemicals listed under OSHA's PSM standard.

cover the hazards of the chemicals being used or produced. The Material Data Safety Sheet, or MSDS, is probably the best source of information. Along with the MSDS, information is also required on the process technology. Such items as flow diagrams, safe operating limits of temperature, flow rate and pressure should be provided. Further, piping and instrumentation diagrams, electrical classification, design codes, and standards should all be available to the plant operators.

In order to identify the potential hazards in your chemical processes, a *Hazard Analysis (HA)* must be conducted. The plant should set up a team to conduct plant investigations and evaluations of processes that might be dangerous. This hazard analysis team should focus on the location of each process area, the hazards of the process, the probable outcome if controls should fail, and the possibility of human error. Additionally, it is a good idea to look over the plant's historical record concerning previous incidents – those that caused or could have caused catastrophic results. When the analysis is completed and recommendations have been issued to management for resolving the safety issues identified, it becomes a permanent plant record that should be updated every five years.

PSM requires wastewater treatment plants to develop written Standard Operating Procedures (SOPs). SOPs should describe the steps of each phase of the operation, the operating limits of the process, how to avoid exceeding the limits, safety systems and how they operate, and hazard control for confined space entry and line-breaking activities.

Before work is begun on any new wastewater treatment process, workers must be trained. Workers should be thoroughly trained on how to conduct each process step as safely as possible. Specifically, training should focus on operating procedures, on process-specific safety and health hazards, on emergency operations and shutdowns, and on applicable safe work practices. Followup refresher training should be completed every three years.

One of the interesting requirements of the PSM Standard deals with outside contractors. Contract workers who work on the wastewater treatment facility are exposed to the same hazardous processes as are plant personnel. PSM spells out the duties of the host (employing facility) and the contractor. Some of the specific duties of the host include:

- informing the contractor of the potential hazards
- explaining the plant's Emergency Response Procedures
- evaluating the contractor safety record and programs (*NOTE:* OSHA requires plants that employ outside contractors to perform work in or around a covered process to be screened to ensure that the contractor can complete assigned work without compromising worker safety and health.)
- conducting safety audits of the construction site

Some of the contractor's specific duties include:

- ensuring that contract employees follow facility safety rules
- informing contract workers of the process hazards and the plant's Emergency Response Plan
- ensuring that each contract worker has been properly trained to perform his or her job safely

PSM contains several other requirements. It requires the use of *Hotwork Permits,* for example. Whenever hotwork such as welding, grinding, brazing, or burning work is performed on or near a covered process, the danger exists that heat generated by the hotwork will change the process. This change might be extremely dangerous. Welding on a digester methane line, for example, could bring about changes to the process that could destroy the entire site and many workers. The Hotwork Permit should verify that necessary fire-prevention measures have been taken. The typical Hotwork Permit gives the dates that are authorized for hotwork and identifies the object on which the hotwork will be performed. Figure 3.32 shows a typical Hotwork Permit.

When working with outside contractors it is important to get off on the right foot before construction begins. As stated previously, PSM requires the host to inform the outside contractor of the hazards and/or potential hazards of the plant site. This can be accomplished by "briefing" the contractor senior personnel prior to the start of construction.

An effective means of "briefing" the contractor is to conduct this "briefing" in a formal manner. A key point to remember is that your presentation should be a "briefing" only. Never turn your briefing into a *training* session. When you train personnel, you become liable for their actions. Figure 3.33 shows an attendance roster with "briefing" information and a disclaimer that is used to "brief" outside contractors.

In addition to the subject matter and disclaimer contained in Figure 3.33, it is wise to incorporate another form into your PSM procedure, developed for the short-term contractor. For example, if your plant contracts with an outside contractor to replace a door or a window and this work will take less than 24 hours, you must still brief the contractor on the hazards and/or potential hazards at your site.

The form shown in Figure 3.34 accomplishes the verification requirement of the PSM Standard. In addition to reading and signing this form, the outside contractor is briefed on the plant's process hazards (e.g., methane, hydrogen sulfide, chlorine, anhydrous ammonia, etc).

Only some of the requirements under OSHA's Process Safety Management Standard have been covered here. The wastewater treatment safety official should refer to the requirements in 29 CFR 1910.119 for further guid-

HAMPTON ROADS SANITATION DISTRICT
HOT WORK PERMIT
for compliance with
(OSHA 1910.119 – PROCESS SAFETY MANAGEMENT STANDARD)

SECTION 1 (Please Print)

WORK DESCRIPTION	TIME:	BEGIN	END

TOOLS TO BE USED/SPECIAL HAZARDS	DATE:

PERMIT ISSUED TO (NAME/COMPANY)	PERMIT ISSUED BY (NAME)

SECTION 2

To be verified by HRSD site supervisor of area where work is to be performed

ITEM	YES	NO	N/A	COMMENTS	ITEM	YES	NO	N/A	COMMENTS
Lines/Tanks Washed					Interfacing Areas Notified				
Lines/Tanks Drained					Extinguisher Present				
Lines/Tanks Pressure Vented					Confined Space				
Lines Blinded/ Disconnected					Oxygen Level*				Level:
Valves Off Locked/Tagged					L. E. L. *				Level:
Power Off Locked Tagged					Fire Watch	Name/Company:			

I certify all the items above have been completed and hereby authorize this permit.

HRSD or CONTRACTOR

Atmosphere Tester Signature_____ Site Supervisor's Initials_____
(Asterisk items only)

SECTION 3

To be completed by Maintenance or Contractor personnel.

ITEM	YES	NO	N/A	COMMENTS	ITEM	YES	NO	N/A	COMMENTS
Lines Blinded/ Disconnected					Glasses/Gloves				
Valves Off Locked/Tagged					Protective Cloth				
Power Off Locked Tagged					Area Roped/ Barricade/				
Air Mask					Signs In Place				
					Fire Watch Present				
Air Bottles Checked					Screens & Curtains				

I certify all the items above have been completed and hereby authorize this permit.

Maintenance/Contractor signature_____

White Tag copy - Display at work site   Pink copy - Supervisor's Log copy

FIGURE 3.32. Hot work permit (*Source:* Hampton Roads Sanitation District; used by permission).

108

In accordance with the recordkeeping and training requirements under OSHA 29 CFR 1910.119 Standard, I have received a safety brief (in no way should this HRSD brief be construed by anyone as taking the place of required contractor employee safety training) from HRSD Safety Division personnel covering Hazard Communication, Lockout/Tagout, Confined Space, and Personal Protective Equipment and Safety Rule procedures utilized by HRSD personnel. I further understand that HRSD expects outside contractors, including sub-contractors, suppliers, agents and employees of such, to perform construction activities under OSHA required guidelines and HRSD procedures. It is understood that all information regarding the Hazards, Safety procedures, Lockout/Tagout procedures, Hot Work Permit procedures and systems operations shall be disseminated to all persons, sub-contractors and agents employed either directly or indirectly by us. I agree to submit documentation as requested by HRSD to verify that my employees, sub-contractors and any other parties working directly or indirectly for me are conveyed the information required.

Signature: Print Name: Company:

FIGURE 3.33. Outside contractor briefing roster (*Source:* Hampton Roads Sanitation District; used by permission).

OUTSIDE CONTRACTOR LOG

By signing and dating this form I acknowledge that I fully understand that this HRSD facility fully complies with applicable OSHA/EPA standards (in particular with the 29 CFR 1910.119 Process Safety Management Standards, Hazardous Chemicals in process treatment) and expects all outside contractors to abide by the same requirements.

During your stay at this facility if you should require assistance or the need arises for you to visit other locations on the facility, please notify plant management.

SIGNATURE

CONTRACTOR NAME	PERSON IN-RESPONSIBLE CHARGE	DATE

FIGURE 3.34. Outside contractor log (*Source:* Hampton Roads Sanitation District; used by permission).

109

ance. Smaller wastewater facilities with limited resources may consider decreasing associated risk by reducing their inventory of hazardous materials. The important thing to remember is that the potential for danger is present at your plant every day. By following the guidelines in the Process Safety Management Standard you can contribute to the protection of your workers, contractors, and the local community.

MACHINE GUARDING (29 CFR 1910.212)

The wastewater treatment process would quickly come to a halt if it were not for the machines that provide the motive force to move the wastestream through the process. From sewage lines to interceptor lines through pumping stations and the pressurizing process, the influent is literally forced into the treatment plant by machines (gravity flow systems excepted).

But the work of machines does not stop here. Inside the plant, a variety of other machines and machine-operated devices screen the flow, remove grit, and then provide the force needed to push flow into primary and then secondary treatment.

The need for additional machines continues as the wastestream enters and leaves the various treatment processes. For example, during the treatment process huge motor-driven blowers are used to aerate the flow. Later, when liquids and solids are separated, both wastestreams continue to move along, powered by machines.

It should be evident from the preceding discussion that the wastewater treatment process uses several different machines. When properly maintained and operated, these machines make the operator's job easier, his/her performance more efficient, and the workplace safer.

Worker safety is enhanced by machines, but only if the machine itself is safe. The basic purpose of machine guarding is to prevent contact of the human body with dangerous parts of machines. When body parts such as arms, fingers, and hair make contact with moving machinery, the result can be disastrous and sometimes fatal. Some of the most gruesome accident investigations involve body part amputations caused by contact with unguarded moving machinery.

The plant safety official must quickly become familiar with his/her plant's moving machinery. Moreover, the plant safety official must be familiar with the methods of machine guarding. OSHA has published a machine guarding reference source (publication no. 3067), *Concepts and Techniques of Machine Safeguarding*. This small booklet is highly recommended. It can be obtained from the Superintendent of Documents, U.S. Government Printing Office, Washington, DC 20402. *NOTE:* The intent of the information that follows is to familiarize you with the hazards of unguarded machines, common safeguarding methods, and the safeguarding of machines.

Unguarded machines can present different types of hazards. One type of hazard is *mechanical,* which includes hazards arising from the motions or operations of machines. Examples of these machine motions and actions are listed as follows:

- rotating machines
- reciprocating motions
- transverse motions
- cutting actions
- punching, shearing, and bending actions

The other type of hazard is one unrelated to the movement of the machine (nonmechanical hazards). *Nonmechanical hazards* include electrical power sources, high-pressure systems, chemical emissions, noise, and contact with flying objects (e.g., flying pieces of metal from cutting operations).

Moving machinery is a general inspection item common to all safety audits in wastewater treatment facilities. The safety auditor should ensure that moving parts of machines and points of operation are guarded adequately—even if the machine is normally inaccessible.

Some safety officials feel that inspecting to ensure that machine guards are in place is the extent of their responsibility in this vital area. Nothing could be further from the truth. Experience has shown that ensuring machine guards are in place is only one of many items that need to be checked. As a case in point, consider these additional inspection items and the questions they pose:

- Is there adequate supervision to ensure that workers are following safe machine operation procedures?
- Is there a training program to instruct workers on safe methods of machine operation?
- Is a regular program of safety inspection of machinery and equipment conducted by each work center?
- Is all machinery and equipment kept clean and properly maintained?
- Is equipment and machinery securely placed and anchored?
- Can electric controls for each machine be locked out for maintenance or repair?
- Are all emergency stop buttons colored red?
- Are all moving gears and chains properly guarded?
- Are machine guards secure?

Machine-Guarding Methods

The plant safety official must be familiar with common machine-guarding methods. These include guards, devices, distance and location, and feeding and ejection methods.

Guards can be of several types (Figure 3.35), including fixed, interlocked, adjustable, and self-adjusting. Machine moving parts that are not provided with a manufacturer's guard sometimes must still be guarded. For example, in pumping stations it is not unusual to have an electric motor on the top floor of a multifloor station. The motor drives the pump by exerting force along a very long shaft that runs the pump located on a lower level of the station. Sometimes these long rotating shafts are not protected against human body contact. In this particular case, the remedial action might be accomplished by manufacturing a shaft cage made from flexible but strong wire-mesh screen. After determining the proper size of the wrap-around wire-screen cage, the assembly is then fitted around the entire exposed shaft length and securely fastened in place. It is important to ensure that the freewheeling shaft will make no contact with the fixed cage. Such a wire-screen shaft guard is a simple device that has an important safety purpose.

Fashioning one's own safety guards has the advantage of bringing the workers into the process. That is, when workers are assigned to manufacture their own machinery guards, they often profit from the experience in that they gain understanding of the guard's intended purpose (see Figure 3.35).

Once machinery guards are inspected to ensure that they are in place, two other major items also require the plant safety official's attention on a continuing basis: worker training and clothing. Workers should receive training on each machine and the purpose of each machine guard before being assigned to operate the machine. Periodic refresher training should also be provided. Machine guard training should emphasize the potential hazards of clothing and machines. Oversized clothing can easily catch on machine parts. Moreover, employees should be instructed to wear chemical-resistant clothing and safety shoes if the machine creates an exposure to oils or chemicals.

The last thing about safeguarding machines to be covered here deals with the machine that is not in operation. Workers should not be fooled into performing unsafe actions around machines that are at rest. Several types of machines run in cycles. For example, plant air compressors run until the accumulator is filled and the system is pressurized to the design limit. When the system is fully charged with air, the compressor normally shuts down. When the system line pressure falls to a predetermined level, the compressor restarts—cycles. Figure 3.36 shows a sign/label that should be installed at or in close proximity to machines that cycle.

FIGURE 3.35. Different types of machinery guards.

113

THIS EQUIPMENT

STARTS

AUTOMATICALLY

FIGURE 3.36. Typical machine warning sign/label.

CHAINS, SLINGS, AND ROPES

This section describes the various types of hoisting apparatus used in wastewater treatment. Specifically, hoisting apparatus in the form of chains, slings, and ropes will be addressed. Moreover, safety considerations and inspection of these devices will be discussed. OSHA has published two excellent references *Sling Safety* and *Materials Handling and Storage* (Publications 3072 and 2236). They can be obtained from the Superintendent of Documents, U.S. Government Printing Office, Washington, DC 20402.

During safety audits, particular attention must be paid to hoisting apparatus. Considering the large number of different types of hoisting apparatus that are used in the wastewater industry, it is not surprising that these devices seem to be just about everywhere. Moreover, with their large numbers and frequent use, it is not surprising that hoisting devices play a large role in on-the-job injuries.

Wastewater personnel handle materials on a daily basis. Sometimes these materials need to be moved from one location to another. Chlorine 1-ton cylinders must be changed out and moved, for example, in order to keep the plant disinfection process in operation. Obviously, moving 1-ton cylinders of chlorine or other chemical would not be possible without some type of hoisting apparatus.

The hoisting devices that seem to require the most attention are chains, slings, and ropes, which are commonly used between cranes and hoists and the load so that it can be lifted and moved to the desired location. When using chains, slings, and ropes in the hoisting process, workers must visually inspect each device before and during operation. Damaged or defective chains, slings, and ropes must be removed from service.

To reduce the chances of worker injury and property damage, the following general rules should be observed:

(1) Slings must not be shortened with knots or bolts or other makeshift devices.

(2) Never attempt lifts that exceed the rated load capacity of the chain, sling, or rope.

(3) Sling legs that have been kinked must not be used.

(4) Suspended loads must be kept clear of all obstructions, and crane operators should avoid sudden starts and stops when moving suspended loads.

(5) Natural and synthetic fiber rope slings should be immediately removed from service if any of the following conditions are present:
  • abnormal wear
  • broken or cut fibers
  • powdered fibers between strands
  • variations in the size or roundness of strands
  • discoloration or rotting
  • distortion of hardware in the sling

(6) Use of repaired or reconditioned fiber rope should be strictly prohibited.

(7) Synthetic web slings should be immediately removed from service if any of the following conditions are present:
  • melting or charring of any part of the sling surface
  • snags, punctures, tears, or cuts
  • broken or worn stitches
  • distortion of fittings
  • acid or caustic burns

(8) Sling legs are kinked.

(9) Slings should be securely attached to their loads.

(10) Hands and fingers should not be placed between the sling and its load while the sling is being tightened around the load.

(11) Shock loading is dangerous and must be avoided.

Wire rope is the most widely used type of rope sling in industry. *Wire rope* is defined as a twisted bundle of cold drawn steel wires. It is usually composed of wires, strands, and a core. The wires can be any number of grades depending on the characteristics desired. OSHA has listed several precautions for wire rope use. The two major precautions that the plant safety official needs to be aware of and enforce are listed as follows:

(1) Wire rope slings shall not be used with loads in excess of their rated capacities.

(2) Wire rope slings shall be visually inspected at the start of each shift, before each use, and on regular intervals based on conditions of use. If damaged or defective, the sling shall be immediately removed from service. The following are some items to look for when inspecting a wire rope sling:

- ten randomly distributed broken wires in one rope lay, or five broken wires in one strand in one rope lay (*lay* is defined as wires laid together in various arrangements having a definite pitch (lay) to form a strand)
- wear or scraping of one-third the original diameter of outside individual wires
- kinking, crushing, bird caging, or any other damage resulting in the distortion of the wire rope structure
- evidence of heat damage
- end attachments that are cracked, deformed, or worn
- corrosion of the rope or end attachments

Chain made of stainless steel, monel, bronze, and other metals is commonly used for lifting slings. Chain not specifically designed for use in slings should not be used for load lifting. OSHA requires that alloy steel chain slings have permanently affixed and durable identification stating size, grade, rated capacity, and reach.

Chains utilize hooks, rings, or other couplings to allow for fashioning of the load lifting device. These attachments must have a rated capacity at least equal to that of the chain elements. When inspecting chain slings, inspect all parts as follows:

- for evidence of corrosion or other wear
- for evidence of faulty welds
- for cracking of the links or other parts
- for evidence of the chain links stretching or bending
- for evidence of the hooks stretching beyond 15% of the normal throat openings or twisting beyond ten degrees from the plane of the unbent hook

While conducting the plant safety audit, you must look closely at all devices used in the hoisting process. It is difficult to conduct a safety audit without finding several hoisting devices that are unsafe for use. Always confiscate these devices and destroy them. If you leave them in the workplace, sooner or later they will be used again.

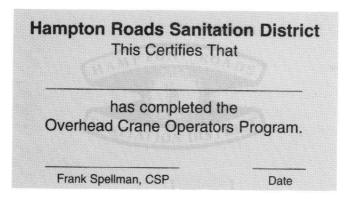

FIGURE 3.37. Overhead crane operator's certification card (*Source:* Hampton Roads Sanitation District; used by permission).

During your safety audit, there are additional things that you should look for concerning hoisting practices and equipment. First, insist that workers accept the responsibility for their own actions. Second, the workers must be competent and careful. Finally, workers who are involved in hoisting activities in your plant need to be trained on chain, sling, and rope safety; also, as shown in Figure 3.37, document the training.

BLOODBORNE PATHOGENS (29 CFR 1910.1030)

Although scientific research has determined (at present) that HIV and other bloodborne pathogens are not found in the wastewater stream (except in strictly controlled laboratory conditions), the Center for Disease Control (CDC) does warn that persons who provide emergency first aid could become contaminated. Thus, if your facility requires employees trained in first aid to render medical assistance as part of their job activities, your facility is covered under this standard. The major point to get across to all workers is that if they render any kind of first aid whereby the rescuer can/could be exposed to another person's body fluids, then care and caution must be exercised.

It is a good idea to equip all wastewater facilities with first aid kits that are designed to protect against bloodborne pathogens. Such kits are equipped with the following:

(1) Rescue barrier mask—prevents mouth-to-mouth contact
(2) Alcohol cleansing wipes—for cleanup
(3) Latex gloves—prevents hand contact with body fluids
(4) Safety goggles—prevents body fluids from entering eyes
(5) Biohazard bag—for disposal of cleanup materials

Training should be provided to each worker on the dangers of bloodborne pathogens. Careful attention to personal hygiene habits should be stressed. Workers should be informed that handwashing is one of their best defenses against spreading infection, including HIV. Worker awareness is the key to complying with this standard.

In summary, this chapter has reviewed many of the safety programs that are required by OSHA and other regulators to be used in industry and at wastewater treatment facilities. Several other safety requirements apply, depending on the size and nature of your operation. Safety programs that deal with safe forklift driving, material handling, and first aid procedures, for example, are important. If you are not sure of your requirements in this area, contact your local branch of federal or state OSHA for guidance.

CHAPTER 4

Recordkeeping

Personnel employed in the wastewater industry quickly learn about recordkeeping requirements. For example, wastewater treatment facilities that discharge to state waters must have a discharge permit issued by the state Water Control Board or other official agency, as the National Pollutant Discharge Elimination Systems (NPDES) permit.

The NPDES permit has very specific and detailed requirements such as recording monitoring information, instrument calibration and maintenance, certain reports required, and data used to complete the permit application. All records must be kept at least three years (longer if requested).

Because of the NPDES permit recordkeeping requirements, the wastewater treatment plant operator soon learns that recordkeeping is an important part of his/her job. How important is this recordkeeping? The best way to answer this question is to consider the average routine within a wastewater treatment plant.

You might be able to measure the pulse of the treatment operation by observing the strict attention that is given to keeping the plant operating log current. Among managers, operators, and assistants the normal every-day—every hour—topic of discussion seems to relate to the plant's need to "make permit" for the month. In other words, plant personnel are geared to constantly control the plant process to keep within specific permit guidelines or limits. An integral and critical part of "making permit" is data recording in the plant's operating log.

The plant's on-going effort to "make permit" and the never-ending record-keeping requirement brings up another point; that is, if the plant does as good a job in maintaining its safety and health records as it does with its permit records, then it will be "making permit" on a big-time scale.

How important are safety and health records? Under the OSHA Act, certain recordkeeping requirements have been mandated. Moreover, DOT and EPA also have certain recordkeeping requirements.

Safety and health records are vital for several reasons. First, as previously stated, they are kept because they are required. Second, without records, plant management has no way of knowing how the plant is performing. Without safety and health records on accidents and injuries, management may be unable to remedy hazards and provide prevention methods. Moreover, without the legally required records on safety and health, management may find itself subject to civil penalties. For example, if plant management fails to maintain employee health records (e.g., audiometric test results), the result may be a liability suit for a former employee's hearing problem, even years after the alleged exposure took place.

OSHA REQUIRED RECORDS

OSHA-200 Record Form

The wastewater industry comes under OSHA's recordkeeping requirements. The purpose of this requirement is to enable OSHA to develop statistical information concerning illnesses and injuries.

First, OSHA requires employers to complete and maintain the OSHA Form 200 to record and classify occupational on-the-job injuries and illnesses. Employers may use the form provided by OSHA (see Figure 4.1) or an equivalent form including computer-generated forms.

The plant employee responsible for maintaining the 200 log must ensure that it is kept accurate and up-to-date, and that over- or underreporting of accidents and illness is avoided. Moreover, the Summary of Occupational Injuries and Illnesses must be completed at the end of each calendar year and posted on employee bulletin boards where they are readily visible to all workers. The posting must occur no later than February 1 of the next year and must remain posted until March 1 (see Figure 4.1).

Medical Records

OSHA also requires employers to maintain accurate medical records. For example, if your plant uses respirators for confined space entry or other purposes, you must have and maintain a medical surveillance program that includes individual employee medical records. In addition, many wastewater facilities require physical examination before placement, on an annual basis, and on termination. Medical records must be maintained with great accuracy and confidentiality.

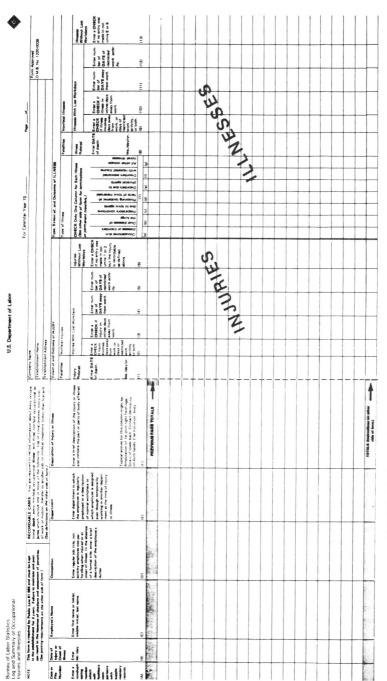

FIGURE 4.1. Log and summary of occupational injuries and illnesses (OSHA No. 200).

121

Training and Safety Audit Records

When an OSHA inspector arrives at your plant site, it is important that you have accurate and current training records for his/her review. Records of safety training should contain the following elements:

- employee's name
- date training was given
- subject matter
- length of training
- instructor's name and title

In addition to training records, the plant must maintain records of safety audits. Safety audits should include the following recorded information:

- name and title of inspector
- audit date
- audit schedule
- area audited
- list of discrepancies observed

Management personnel can conduct all the safety training and auditing they want to do. However, if they do not record the training or the auditing, then, in the eyes of OSHA, it did not occur.

EPA REQUIRED RECORDS

The wastewater treatment industry not only treats waste, it also produces waste. Sometimes it produces hazardous waste. Whenever this is the case, the EPA gets involved and also requires recordkeeping. Through its Superfund Amendment and Reauthorization Act (SARA), the EPA requires industries that handle, store, or generate hazardous waste to comply with reporting and recordkeeping requirements.

One such requirement has to do with reporting and recording hazardous materials incidents. For example, in a wastewater treatment plant that uses chlorine in the disinfection process, the possibility of a chlorine leak is always possible. If such a leak occurs and if more than 10 pounds of chlorine is released into the environment, the EPA must be notified.

SARA also requires the plant to maintain a Hazardous Chemical Inventory. This inventory is another recordkeeping requirement. It must be made available to the Local Emergency Planning Committee (LEPC), State Emergency Response Commission (SERC), Local Fire Protection Agency, and any local citizen of the community who might ask to see it.

DEPARTMENT OF TRANSPORTATION (DOT) RECORDS

If your wastewater treatment facility transports hazardous materials above the Reportable Quantity (RQ) (e.g., one-ton cylinder of chlorine; RQ for chlorine is 10 pounds) on public highways, the DOT requires you to maintain specific records. The three types of records to be maintained include:

- driver records
- accident reports
- maintenance and compliance records

SUMMARY

This chapter has presented a brief summary of the recordkeeping requirements connected with safety and health compliance. The importance of maintaining records cannot be over-emphasized. The wastewater treatment plant that keeps accurate and current NPDES-required reports stands a good chance of "making permit" each month. In addition, the plant that maintains accurate and current safety and health records stands a good chance of avoiding an official citation or a liability suit.

CHAPTER 5

Safety Training

Throughout this text the major point has been to emphasize and reemphasize the importance of *employee safety training*. This emphasis has been for good reason. For without a doubt, providing routine safety training for workers is one of the most important job duties of the safety official in the wastewater industry. Indeed, most managers know the importance of safety training, but is not well known that specific training requirements are detailed in OSHA, DOT, and EPA regulations. Under OSHA regulations, for example, it is stated or implied that the employer is responsible for providing training and knowledge to the worker. Moreover, employees are to be apprised of all hazards to which they are exposed, along with relevant symptoms, appropriate emergency treatment, and proper conditions and precautions of safe use or exposure.

To this point in the text, several OSHA safety and health standards or programs have been featured. Employers must comply with these standards and must also require workers to comply. More than 100 OSHA, DOT and EPA safety and health regulations contain training requirements.

It is interesting to note that although OSHA requires training, it does not always specify what is required of the employer or entity that provides the training. However, information and instruction on safety and health issues in the workplace are foundational to building a viable organizational safety program. Workers cannot be expected to perform their assigned tasks safely unless they are aware of the hazards or the potential hazards involved with each job assignment.

For years safety professionals have spoken about the "three E's" of safety—Engineering, Enforcement, and Education. While it is true that the best solution to controlling any hazard is to engineer out the problem, and while it is also true that enforcement is critical to executing proper safe work practices, it is also true that the safety education aspect is important.

Education in the form of providing information and training is one of the most vital elements in safety simply because workers cannot be expected to comply with safe work practices unless they have been informed of and trained on the proper procedures.

Experts in the safety field have differing points of view on this topic. One element points out that safety is not a behavioral issue, instead, it is technical—meaning that safety can be accomplished by engineering out the hazard. It should be remembered, however, that if there is a possibility for something to go wrong, workers will find the way to make it happen. Even workplaces that have state-of-the-art engineering safety devices and strong enforcement programs do not always have effective hazard control programs unless the workers and supervisors understand the hazards and the potential for hazards that arise from not observing routine safe work practices. A worker's work routine cannot be engineered—workers are not robots.

Worker safety training should begin right after the employee is hired. New Employee Safety Orientation Training programs can be effective if correctly structured and presented early in the worker's tenure. Figure 5.1 shows an outline of New Employee Safety Orientation training that has been used for several years. This training procedure has proven to be quite effective by getting the new worker off to a right start. The training also satisfies several OSHA requirements. For example, OSHA requires that workers be trained on the Hazard Communication Standard prior to beginning work with or around hazardous materials.

Before sending new employees to safety orientation training, it must be determined who needs what training. The plant safety official in conjunction with the personnel manager, work center supervisors, and safety council should determine specific safety training for each job classification. This can be accomplished by conducting a "needs assessment" (sometimes called "needs analysis"). The idea behind the needs assessment is to ensure that job classifications requiring confined space training, for example, receive this critical training. At the same time, the needs assessment also functions to ensure that the plant clerk who needs hazard communication training but who does not enter confined spaces, for example, does not receive confined space training.

Depending upon the organization's turnover rate, the frequency of conducting New Employee Safety Orientation Training (Figure 5.1) can vary from presenting a session each week, every other week, once a month, or as required. As stated previously, it is important that good records of worker safety training are kept.

New employee Safety Orientation training will be conducted at the Nansemond Treatment Plant on Tuesday, July 25, 1995, 0700–1530. Supervisors should arrange to send those employees who have not yet attended one of these sessions. Lead Operator candidates and others who need their PQS forms for Safety signed by Safety Division are also encouraged to attend.

Approximate times are listed for each topic scheduled to allow Supervisors to plan for their personnel who may need refresher training in specific areas.

0700–0815—Employee ID cards and Personnel Benefit Briefing.
                District Substance Abuse Policy.

Safety Agenda:

(1) HAZCOM 0830–0945
(2) NFPA labeling and DOT packaging 0830–0930
(3) Hotwork Permits 0945
(4) Lockout/tagout 0950–1040
(5) Respiratory protection—SCBA 1230–1330
(6) Confined space entry 1345–1445
(7) HM-181 0830–0930
(8) Fire awareness/extinguisher training 1115–1130
(9) Injury reporting—Supervisor's 1st Report of Injury Form; procedure for reporting unsafe conditions. 1050–1100
(10) Basic first aid 1450–1530
(11) Power tool/grinding wheel safety 1230
(12) Back injury prevention training 1450–1530
(13) Chlorine safety 1105–1115
(14) Hearing conservation 1230
(15) Personnel protective equipment 0830
(16) H2S 0830–0930
(17) Fall protection/ladder safety 1120

Note: Personnel Division will issue new employee ID cards during this session. Moreover, the Personnel Specialist will make an important and beneficial presentation concerning employee rights and benefits.

FIGURE 5.1. New employee safety orientation outline (*Source*: Hampton Roads Sanitation District; used by permission).

In addition to New Employee Safety Orientation Training that is presented to all new employees, another way to verify new employee safety knowledge is a system known as *Personnel Qualification Standard (PQS)* for plant safety. PQS is a term that comes from the United States Navy. The Navy has been using PQS to qualify its personnel for several years. Like the Navy's PQS program, wastewater treatment plant PQS has proven to be an excellent training instrument. Not only does it provide a guide of what is to be learned, it also provides documentation to show that the training was actually completed. Additionally, this wastewater PQS system is suitable for water treatment plants and for just about any other industrial application.

Robert Rutherford, Plant Manager of the HRSD James River Treatment Plant, Newport News, Virginia, and the author have developed a wastewater treatment plant safety PQS that is printed in booklet form. The PQS booklet is an outline of training requirements that must be satisfied prior to the new employee being put to work at a treatment plant. The average time of completion of this PQS is 8 hours; 8 hours that is well invested.

Rutherford ran a pilot study for PQS at his James River Treatment Plant for one year. Results from the pilot study indicated that PQS works; the on-the-job injury rate decreased by 75%. After the success of the pilot study, PQS was implemented at eight other wastewater treatment plants and one compost facility at Hampton Roads Sanitation District. After five years of use, HRSD's PQS was reevaluated. Not surprisingly, the significant result of PQS was a 68% decrease in the rate of on-the-job injuries. Prior to the implementation of PQS at these same plants (prior to 1990), the on-the-job injury rate averaged at least five injuries per month. The frequency average of on-the-job injuries to date is less than one per month.

When attempting to implement a PQS program at your plant, keep in mind that several requirements must be completed prior to the new employee being fully safety PQS qualified. The PQS booklet guides the new employee in learning about the treatment plant safety organization. Moreover, the new employee is provided with a plant safety tour where he/she receives a site map. The site map shows all buildings, process units, and chemical storage locations. All significant items on the site map are *not labeled*. It is the new worker's responsibility to tour the plant site (in the company of the chief operator) and to fill in the names of all buildings, process unit machinery, and chemical storage locations. In addition to naming various components within the plant site, the worker must also note on the plant site map where all plant first aid kits, fire extinguishers, fire hoses and standpipes, emergency equipment storage lockers, and emergency evacuation routes are located.

After filling in the appropriate entries on the site map, the new worker is required to fill in blank MSDS forms for each of the plant's hazardous materials (e.g., chlorine, sulfur dioxide, ammonia, etc.). Along with filling out the MSDS forms, the worker must locate and understand the meaning of all labels and warnings that are attached to chemical storage tanks or outside chemical buildings.

The new worker then moves on in the qualification process and is shown all the plant's confined spaces. The tour guide explains the plant's confined space entry program and the dangers involved with confined space entry.

PERSONNEL QUALIFICATION RECORD

EMPLOYEE_____   HIRING DATE_____

PLANT SAFETY INDOCTRINATION

1. Treatment Plant Safety
    Organization
    _____   _____   _____   _____
    Employee   Date   Instructor   Date
    Signature   Initials

2. Plant Safety Tour
    _____   _____   _____   _____
    Employee   Date   Instructor   Date
    Signature   Initials

3. Hazard Communication
    _____   _____   _____   _____
    Employee   Date   Instructor   Date
    Signature   Initials

4. Confined Space
    _____   _____   _____   _____
    Employee   Date   Instructor   Date
    Signature   Initials

5. Breathing Apparatus
    _____   _____   _____   _____
    Employee   Date   Instructor   Date
    Signature   Initials

6. Personal Protective
    Equipment
    _____   _____   _____   _____
    Employee   Date   Instructor   Date
    Signature   Initials

7. Safety Rules
    _____   _____   _____   _____
    Employee   Date   Instructor   Date
    Signature   Initials

8. Lock-Out/Tag Out
    _____   _____   _____   _____
    Employee   Date   Instructor   Date
    Signature   Initials

9. Hand Tool, Powered
    Hand Tool, and
    Ladder Safety
    _____   _____   _____   _____
    Employee   Date   Instructor   Date
    Signature   Initials

FIGURE 5.2. New employee qualification record (*Source*: Hampton Roads Sanitation District; used by permission).

129

The new employee is also shown a confined space permit and the tour guide explains the importance of the form.

During the remainder of the tour, the new worker is shown all the hazards within the plant site. In addition, each of the other plant safety programs like lockout/tagout, personal protective equipment requirements, safety rules, respirator storage and respiratory protection requirements are discussed. Upon completion of the site tour, the chief operator initials and dates those areas the new employee has completed.

Upon completion of the PQS booklet and with all knowledge areas properly initialed and dated, the new worker is interviewed by the plant superintendent to ensure that he/she has been informed about all pertinent safety concerns at the plant site. If the plant superintendent feels comfortable with the new worker's level of safety knowledge, the new worker is PQS qualified (superintendent signs) and is turned over to a plant operator to begin his/her wastewater treatment on-the-job training process.

The benefits of using safety training PQS or some other training system or technique for new workers cannot be overemphasized. By utilizing such a safety indoctrination process, supervisors are ensured of training their personnel and creating an accurate record of the initial safety training. Experience has shown that this PQS system is dynamic, constantly changing, and constantly improving. Figures 5.2 and 5.3 show recordkeeping forms used for new employee safety training PQS.

PERSONNEL QUALIFICATION STANDARD
FOR
NEW EMPLOYEE SAFETY TRAINING

PERSONNEL EMPLOYEE SAFETY TRAINING (Cont'd)

Plant Safety Indoctrination Review   _____   _____
                                      Plant Superintendent   Date

District Safety Orientation   _____   _____
                                      Safety Manager   Date

Employee's Acknowledgement

I acknowledge that I have received and   _____   _____
fully understand the training provided   Employee   Date
to me according to this Personnel
Qualification record.

FIGURE 5.3. Personnel final qualification form (*Source:* Hampton Roads Sanitation District; used by permission).

Safety begins with awareness, which, in turn, is gained through training and experience. Whatever type of safety training that is decided upon, it is important to assess the program's effectiveness. Workers who attend safety training should be asked their opinions of the training. Through the use of questionnaires, for example, it can be determined whether the approach being used is relevant and appropriate. Supervisors are another good source of information. Since they are in a good position to determine if the safety training is paying off. In addition, improvements should be noticeable in the workplace. Reduced numbers of injuries on-the-job along with improved product quality and worker productivity are excellent indicators of the effectiveness of the plant's safety training program. Finally, as stated previously, proper safety training can be accomplished but if it is not documented — the training was never done.

CHAPTER 6

Safety Equipment

Throughout this text the importance of Personal Protective Equipment (PPE) and training accompanying its use has been pointed out. In addition, OSHA-mandated compliance requirements for most employers in providing employees with approved PPE has been discussed. This chapter takes worker protection beyond the PPE stage.

As pointed out earlier, PPE basically is to be used as a last resort in protecting workers. Work sites should be thoroughly evaluated for hazards and hazards should be removed or engineered out prior to workers being required to preform work in them. Obviously, not all hazards can be eliminated or engineered out of the workplace. Wastewater treatment and collection is one of those industries where hazards cannot be completely eliminated. For example, confined space entry and chemical handling are hazards that are inherent to the industry and cannot be easily removed or done away with. However, performing work in confined spaces and handling chemicals can be made safer if the proper safety equipment is available. Moreover, safety equipment is only as good as the proficiency of the workers who use it.

Here again, the importance of proper safety training cannot be overemphasized. In the following pages some of the specific safety equipment that should be used while performing particular work activities within the wastewater treatment/collection industry is pointed out.

SAFETY EQUIPMENT AND CONFINED SPACE ENTRY

Background

29 CFR Subpart Z—Toxic Substances 1910.1000 Air Contaminants—sets

the Permissible Exposure Limits (PELs) to protect workers from exposure or overexposure to hundreds of airborne contaminants. The "Z Tables" list all the exposure limits for each substance regulated by OSHA. Where the individual standards are written for substances, the "Z Tables" tell you where to look for exposure information within the Code of Federal Regulations.

Air Monitoring

OSHA's 29 CFR 1910.146 – Confined Space Entry – sets guidelines for monitoring confined spaces. For example, prior to entry into a confined space, wastewater treatment facilities and collection systems are required to use air monitors for oxygen, combustible, and toxic gases that are commonly found in confined spaces.

When air-monitoring equipment is used to test the atmosphere of a confined space, it is important to follow the guidelines for testing and recording of results provided on the Confined Space Entry Permit Form. The Confined Space Entry Permit Form (see Figure 3.6) basically lists the steps to be followed in order to effect proper and safe confined space entry.

Along with monitoring the atmosphere of confined spaces for air contaminants and/or lack of oxygen, the Confined Space Entry Permit and the Plant's Confined Space Entry Procedure should point out what type of additional safety equipment is required to protect workers when entry into the confined space is to be made.

CASE STUDY

With regard to air monitoring for confined spaces and fatalities occurring in confined spaces, consider the following example of an incident that actually occurred.

Recently, a construction crew obtained a contract from an industrial complex to clean and preserve the interior surface of a large chemical storage tank. The construction crew was not familiar with OSHA's Confined Space Entry Permit Standard. When the construction workers arrived at the work site, and after having peered inside the tank, they quickly realized from the horrendous stench emanating from the interior of the tank that it might not be a safe environment in which to work. The crew foreman had an idea. He quickly left the tank and his crew and sought out the work site's manager and asked to borrow an air-monitoring device. Reluctantly, the site manager loaned the foreman a portable air-monitoring device. Armed with the air monitor, the foreman went back to the tank and informed his crew that he was going to test the inside atmosphere for contaminants. This is where the first problem arose. The foreman had never used an air monitor before. He did not know how to use it; how to calibrate it.

After about 20 minutes of frustration from not being able to get the monitor to work, the foreman gave up trying. Instead, he directed two of his crew members to enter the tank to determine what equipment would be needed to start the preservation project.

Fifteen minutes after the two crew members had entered the tank, the foreman decided to find out how they were progressing. He walked over to the tank, stuck his head into the entry port and asked how it was going. There was no answer. The foreman asked again. Still no answer.

At this point the foreman became concerned; he panicked. Without thinking clearly, he grabbed a flashlight and entered the tank.

What exactly happened then is unclear. One point is very clear, however. The next morning one of the construction officials showed up at the plant site to see how his crew was doing on the tank job. The construction official knew his men were on the job because the company truck was parked next to the tank.

The construction official could not locate the crew and decided that they must be working inside the tank. When he looked inside the tank he saw a dimly lighted flashlight shining on three limp bodies. His crew was either taking a nap or worse. It had to be worse, he reasoned—because no one could be napping in an environment that smelled that bad.

The construction official feared the worst. He ran into the site office and dialed 911. Later that day the results were known: three dead workers.

This case illustrates that there are lessons to be learned. In the first place, no company official should loan safety equipment to an outside contractor or to any other non-company person. When you loan your equipment, you may also assume liability for whatever occurs. In this case, the loaning of plant equipment was the wrong thing to do. Additionally, when the site official did not show the foreman how to operate and calibrate the air monitor, he assumed liability for the foreman's actions—he bought into the incident and its consequences. In the second place, the construction crew had not been trained on proper confined space entry procedures—if they had, OSHA would not have been there to investigate. More importantly, this incident might not have occurred; the crew would still be alive.

Tripod and Winch, Lifeline and Body Harness Systems

The OSHA Confined Space Standard states that whenever a person enters a vertical permit-required confined space of more than five feet in depth, a mechanical lifting device must be available for rescue purposes. Having the correct safety equipment on hand is not only a matter of compliance—it could literally save workers' lives. *Workers* (in the plural) is stressed here because experience has shown, as the previous case study clearly pointed out, that fatalities involved with confined space entry generally involve more than one victim.

Several types of mechanical lifting devices are available for rescue purposes. The standard portable tripod and winch system used for retrieval is available from several manufacturers (see Figure 6.1). The portable tripod and winch system used for confined space rescue must be equipped with a heavy-duty lifeline capable of retrieving personnel quickly and safely. Along with lifting mechanisms and heavy-duty life lines, an approved *safety harness* must be used. Safety harnesses are particularly valuable when rescuing a helpless or unconscious victim from a confined space. The safety harness allows for safe and quick retrieval in an emergency.

Confined Space Ventilation Systems

When it comes to air quality, OSHA recognizes that all confined spaces can present significant dangers. Handling such situations first involves a careful analysis of the hazards, usually through use of air-monitoring equipment previously mentioned.

Another method can be used to determine whether or not a particular confined space is hazardous. Historical records can be used. Confined space entry permits are generally kept on file for at least one year. Some safety professionals recommend that confined space permits that contain recorded instances of hazardous conditions should be kept on file longer. This is a good idea. If a confined space demonstrates that hazards are present, what is to say that they will not always be present? By compiling data from previous confined space entries, the qualified/competent person will be alerted to the hazards that are possible or probable in the confined space that he/she is attempting to enter. Whether a confined space has a record of hazardous conditions or not, proper air-monitoring procedures must be followed no matter what.

In most cases, an atmospheric hazard can be eliminated or reduced by ventilating the confined space, for example, by using portable blowers. The blower should be used outside the space to be ventilated. Supply/exhaust ducting is attached to a blower and lowered into the vertical or pushed into a horizontal entry confined space to ventilate. Sometimes the use of ducting is difficult and dangerous. Bulky ventilation ducts can compromise safety by restricting worker access to confined spaces. Special ducting attachments have been designed (e.g., the saddle vent) that save time and increase safety. The industrial saddle vent is ideal for wastewater work because it allows for ample airflow, yet takes up only 3″ of manhole clearance, allowing for easy entries and exits (see Figure 6.2).

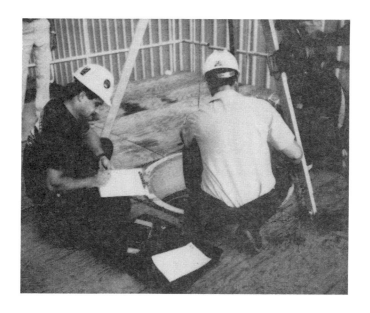

FIGURE 6.1. Tripod and associated safety equipment for confined space rescue.

137

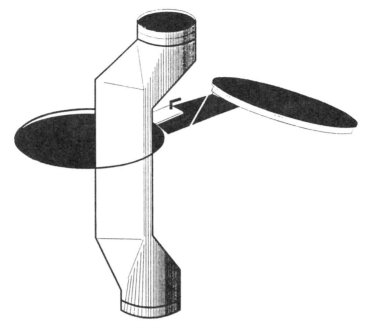

FIGURE 6.2. Saddle vent use for confined spaces.

Portable Lighting Equipment

With the exception of deep pumping station wetwells and some manholes that open into lengthy sewer networks, few environments are more foreboding or darker than the inside of an underground, seemingly endless 60″ diameter interceptor line. Because flammable/explosive gases or vapors might be present in such an environment, vapor-/explosion-proof/confined space hand lamps and ground fault protection are required. Along with using approved incandescent or fluorescent hand lamps, flashlights are often used in confined spaces. When used in confined spaces, flashlights should be the approved gasproof type. Ordinary flashlights are dangerous because they can provide the ignition source to ignite gases and vapors.

Nonsparking Tools

OSHA has specific requirements regarding the use of tools in flammable vapor and combustible residue areas (e.g., some confined spaces). To comply with OSHA and to ensure worker safety, use of nonsparking tools is recommended. Tools made of beryllium and copper eliminate the threat of dangerous sparks.

Warning Signs and Barricades

The last thing confined space entry personnel need is a traffic jam in or around the area in which they are trying to perform their dangerous work. When confined space entry is required, the immediate area surrounding the entry point should be properly barricaded and posted with warning signs (see Figure 6.3). A simple traffic cone and hazard tape around the confined space entry port can be effective in warning personnel about the hazard. If it is not possible to erect some type of safety barricade around the confined space area, it may be necessary to station personnel in the exclusion area to direct traffic away from or around the space. Regardless of the type of barricade and warning system used to protect workers and the public from the confined space, it is important to leave the traffic control and warning devices in place until the work is completed.

CHEMICAL HANDLING AND MISCELLANEOUS SAFETY EQUIPMENT

Emergency Eyewash/Showers

ANSI states that emergency eyewash/showers must be installed within 10 seconds or 100 feet of acid or corrosive hazards. Wastewater treatment facilities usually employ several different acids or corrosives for various purposes. For example, one of the most frequently used corrosive chemicals at wastewater treatment plants is NaOH, sodium hydroxide, or caustic. Caus-

FIGURE 6.3. Confined space entry warning sign.

tic is used in pH, odor control and other applications. Workers who work on caustic systems need to be aware of the associated dangers, the proper spill cleanup procedures, and the locations of emergency equipment such as emergency eyewash and showers (some of this information can be obtained from the MSDS for caustic). The emergency eyewash and showers must be clearly labeled (see Figure 6.4).

The results of safety audits conducted at various wastewater treatment plants commonly indicate, surprisingly, that several emergency eyewash and shower stations were inoperable. Obviously, these inoperable stations were appropriately listed as major discrepancies in the safety audit reports.

Maintaining safety equipment in an operable condition is critical to protecting the worker. Equipment such as emergency eyewash and shower stations is sometimes ignored, only becoming necessary when an accident occurs. Thus, while they are important safety devices, they are often abused through lack of maintenance, inspection, and operation.

In order to prevent emergency eyewash and showers from deteriorating to the point that they are no longer available for emergency use when needed, it is necessary to routinely inspect, test, and repair these devices. Figure 6.5 shows an eyewash inspection tag used at several wastewater treatment facilities. This inspection tag is attached to each emergency eyewash/shower station at the treatment facilities. The tag shown in Figure 6.5 is dated for the current year, listing each month of the year. Inspection of each eyewash/shower is required on a monthly basis. The inspection is indicated by using a hole-punch or some other indicating device on the tag to indicate that the station was inspected for the month designated.

Some wastewater treatment plants employ planned preventive maintenance (PM) programs. Such programs have proven quite effective. Inspection of emergency eyewash and shower systems can easily be added to the plant's PM system. Actual PM practice as just described is quite effective in ensuring that the equipment is kept operational and ready for use in emergency situations.

FIGURE 6.4. Label for emergency eyewash station.

```
┌─────────────────────────────────┐
│              ●                  │
│ EYEWASH INSPECTION TAG          │
│ This eyewash station was        │
│ inspected and certified for     │
│ use on the following dates:     │
│                                 │
│              1995               │
├────────────────┬────────────────┤
│      JAN       │     JULY       │
├────────────────┼────────────────┤
│      FEB       │     AUG        │
├────────────────┼────────────────┤
│      MAR       │     SEPT       │
├────────────────┼────────────────┤
│     APRIL      │     OCT        │
├────────────────┼────────────────┤
│      MAY       │     NOV        │
├────────────────┼────────────────┤
│     JUNE       │     DEC        │
├────────────────┼────────────────┤
│                │                │
└────────────────┴────────────────┘
```

FIGURE 6.5. Eyewash inspection tag (*Source:* Hampton Roads Sanitation District; used by permission).

Ladder Safety

Choosing the right ladder for the job is an important part of working safely. Whether for maintenance or operational reasons, ladders are used extensively in wastewater treatment. For ladder safety, ladders must be sturdy and in good repair. As with inoperable eye wash/showers, treatment plant safety audits normally detect several ladders that are unsafe for use. Ladders are generally ignored until they are needed. When workers determine that they need a ladder, they generally grab the first ladder available and use it. Unfortunately, there is no guarantee that the ladder is safe unless the plant site has an effective ladder inspection program.

Figure 6.6 shows a ladder inspection tag that is similar to the emergency eyewash/shower inspection tag shown in Figure 6.5. During regularly scheduled safety audits, each ladder should be inspected and each inspection tag should be in place with evidence that a current inspection has been completed. In facilities where this tag and safety audit procedure has been used, it has proven very effective. An organized ladder and eye wash/shower inspection process helps ensure that the equipment is ready for use when needed.

Miscellaneous

Several other safety devices or safety equipment should be made

```
                        ●
        LADDER INSPECTION TAG
        This ladder was inspected
        and certified as safe for
        use on the following dates:

                      1995
```

JAN	JULY
FEB	AUG
MAR	SEPT
APRIL	OCT
MAY	NOV
JUNE	DEC

FIGURE 6.6. Ladder inspection tag (*Source:* Hampton Roads Sanitation District; used by permission).

available to plant workers. For example, *First Aid* and *Chemical Burn Kits* should be mounted or placed within close reach and easy access to all employees. These kits must also be inspected frequently to ensure that they are properly stocked with required first aid supplies such as band-aids, dressings, splints, etc. These kits can also easily be added to the plant's PM system.

Emergency Response Storage Lockers must be properly stocked with essential items such as flashlights, helmets, blankets, tools, emergency provisions, first aid supplies, ladders, chain saws, extra batteries, and other devices. Lockers should be inventoried on a routine basis. Further, as part of the plant safety audit, these items should also be spot-checked to make sure they are properly stocked. Whenever an emergency locker is put together on the plant site, it is a good idea to place an inventory list on the door to the locker. With an accessible list of what is required to be stored within the locker, it is easy to spot-check that required materials are in place.

Routinely scheduled safety audits of all plant safety equipment are important in ensuring that vital equipment is available and in good working order. In addition, as with anything else related to safety, training is important. A plant can purchase all the safety equipment it might ever need and feel it is safe. However, unless supervisors and workers are trained properly on the correct use of the safety equipment, even a minor emergency situation could turn into a disaster. To gain a better appreciation for the value of training, consider the following example.

> Even if your plant uses state-of-the-art, OSHA-approved, expensive safety harnesses to protect workers from falls, these devices are worthless unless properly used. For instance, one of your work crews is to perform maintenance inside a vessel that is quite deep; exactly 75 feet deep. The work is to begin at the top inside section of the vessel opening, and the crew is to work its way down inside the vessel as each step of the work is completed. To protect the workers from falling, a sophisticated hoist assembly is rigged and is inspected by an expert prior to use. Later, the crew arrives at the site and dresses in approved safety harnesses. Upon entering the vessel the first worker and his harness is securely attached to the fall arrest system.
>
> There's a problem, however. When the first worker was attached to the hoist system, he was attached by a proper line that was 80 feet long. Remember, the total fall is only 75 feet. Hopefully, someone will discover the error before the worker begins the work.

The point of the preceding example is clear. Even if the best safety equipment is purchased and used, it must be used correctly. To use their equipment correctly, workers must be trained.

In summary, a properly equipped wastewater treatment facility is only as good as the workers who work in it. Along with operable equipment and worker training, it is important that well-established *Safe Work Practices* be employed. Several recommended safe work practices that have been developed for use in wastewater treatment, collection, and laboratory operations will be discussed in the next chapter.

CHAPTER 7

Safe Work Practices for Wastewater Treatment Facilities

Hazards common to wastewater facilities, collection systems, and laboratories include exposure to infectious disease, physical injury, confined spaces, toxic chemical exposure, electrical shock, explosive gas mixtures, noise, oxygen-deficient spaces, and dust, fumes, and mist exposure, to name only a few. The plant safety official can literally be overwhelmed when attempting to reduce the numerous hazards that are present in wastewater work. A tool that the plant safety official can utilize in attempting to put good order and safety practice into typical wastewater work activities is a *Manual of Safe Work Practices*.

Experience has shown that on-the-job injuries are most often caused by unsafe work practices or incorrect procedures coupled with insufficient training and inadequate supervision. In light of this finding, the final chapter of this text is devoted to safe work practices for the wastewater industry. The following listing was developed over several years of investigating on-the-job injuries that occurred in wastewater treatment and collection activities. Many of the practices resulted from observation of good worker actions; others resulted from not-so-good worker actions. While the following safe work practices are not all-inclusive, they do provide a foundation upon which a viable organizational safety program can be built.

The reference or source material for the following safe work practices is of a dual nature. First, each of the work practices is based on actual on-the-job observation of workers performing the work. This is not to say that you must require your workers to perform exactly as outlined in these practices. Instead, each work practice is designed to be used only as a guide from which you can tailor it to your specific operation. Second, technical and expert reference was obtained through various OSHA publications. These publications are listed in the bibliography.

MANUAL OF SAFE WORK PRACTICES
FOR THE WASTEWATER INDUSTRY

Preamble: Personnel should follow adopted safe work practices during the performance of their daily activities. Safe work practices are written guidelines and reference information describing *recommended* practices to follow for performing *general* routine activities in a safe manner.

(1) The supervisor shall be responsible for:
   - training employees on safe work practices
   - enforcing observance of safe work practices
(2) Workers shall be responsible for:
   - familiarizing themselves with safe work practices
   - observing safe work practices in the conduct of activities while employed at this facility
(3) Adopted safe work practices are described in this manual. While not all work activities are described herein, work activities conducted by any employee and supervised by any supervisor should be accomplished in the safest manner possible.
(4) The organizational safety official shall be responsible for ensuring that this manual is kept current and is distributed to all employees.

Work: Aeration Tanks

PRACTICE

(1) Use care and caution when working around aeration tanks.
(2) Ensure that standard handrails and 4-inch toe boards are provided and maintained around open aeration tanks.
(3) Ensure that walkways around open aeration tanks are free of debris, unnecessary equipment, and all possible trip hazards.
(4) Ensure that adequate lighting is provided around aeration tanks. If unsure whether enough lighting is present, contact the safety official.
(5) Ensure that enough ring buoys and line and/or shepherd's crooks are located around open aeration tanks. If unsure about the required number, contact the safety official.
(6) Wear safety harness or life jacket when leaning on or hanging over aeration tank guard rails or safety chains.
(7) Do not sit on aeration tank guard rails or safety chains.

(8) Keep aeration tank walkways free of ice or snow during freezing conditions.

(9) Ensure that aeration tank walkways are free of chemicals, solids, or other substances that may create slippery conditions. Wear safety harness if slippery conditions exist.

(10) Do not remove gratings along aeration tank walkways unless authorized work is being performed. Keep aeration tank walkway hatchway doors closed unless authorized work is being performed. If necessary to remove, put safety cones in place.

(11) Before entering an empty aeration tank, ensure that the atmosphere is tested with a direct reading instrument and that ladders for egress are readily accessible within the tank. Refer to the organization's Confined Space Entry Program for further information.

Work: Digester—Anaerobic

PRACTICE

(1) Use care and caution when working around anaerobic digesters.

(2) Do not smoke, light open flames, or produce sparks in the immediate vicinity of digester tanks, digester control buildings, or other areas housing digester gas handling equipment or pipes.

(3) Post "NO SMOKING" signs at all entrances to digester facilities and at other conspicuous places.

(4) Conduct regularly scheduled inspection and maintenance of digester equipment.

(5) Inspect digester gas piping and gas handling equipment for leaks on a periodic basis. Inspect and maintain digester gas-pressure relief valves and vacuum relief valves on a periodic basis. Inspect and maintain digester gas-collection drip traps and flame arrestors on a periodic basis.

(6) Ensure that both existing and all new equipment installed in digester buildings meet wastewater regulations and national electrical codes.

(7) Follow confined space entry procedure before entering an enclosed digester.

(8) Use an approved direct reading gas-detection device to check for oxygen deficiency and/or combustible gases in an enclosed digester area before entering.

(9) Use only nonsparking tools or explosion-proof portable lighting in a digester area where an explosive atmosphere exists.

Work: Digester Cleaning—Anaerobic

PRACTICE

(1) Use care and caution when working around or inside anaerobic digesters.

(2) Do not smoke, light open flames, or produce sparks in the immediate vicinity of digester tanks, digester control buildings, or other areas housing digester gas-handling equipment or pipes.

(3) Open digester gas-pressure relief valves and vacuum relief valves before beginning to drain a digester tank.

(4) After ensuring gas pressure has been relieved, open all digester roof hatches while draining a digester tank.

(5) Follow all safety precautions listed in the organization's Confined Space Entry Program before entering any tank. A "qualified person" must complete the Confined Space Entry Permit before employees may enter the digester tank. Check and recheck oxygen deficiency and/or combustible gases using a direct reading instrument in a digester tank while draining, cleaning, and repairing the tank each time before entering. Never enter a digester tank that has an oxygen deficiency or excessive combustible gases. Never enter a digester atmosphere alone and unattended. Make sure at least two personnel are present before entering a digester atmosphere: One person to enter the digester atmosphere, the other to remain in the clear to observe in the event of an emergency.

(6) Provide and continue positive ventilation to a digester tank while draining, cleaning, and repairing it.

(7) Enter a digester from the lowest access opening possible.

(8) Observe proper safety precautions when using ladders in a digester tank. Do not carry tools and equipment in the hands while climbing up or down a ladder to a digester tank. Raise or lower tools and equipment into a digester tank using a rope, sling, or bucket.

(9) Use safety harnesses and life lines when entering a digester tank through a vertical entry.

(10) Watch out for slippery footing in a digester tank.

(11) Use only explosion-proof portable lighting in a digester tank.

(12) Use nonsparking tools in a digester tank as appropriate.

(13) Recheck for combustible gases in a digester tank *immediately before* welding or using a cutting torch in the tank. Complete a "Hotwork" Permit.

(14) Do not smoke in a digester tank.

Work: Handling Ash

PRACTICE

(1) Use care and caution when working around ash.

(2) Wear double eye protection when handling ash. Double eye protection consists of goggles and face shield.

(3) Wear a dust mask or half-face air-purifying respirator with prefilter and nuisance dust cartridges when handling ash.

(4) Wear a back support when shoveling or lifting ash containers.

(5) Wear leather work gloves when handling ash.

(6) Ensure that only ash is disposed of in ash disposal containers.

(7) Wet down to minimize dust when conveying or moving ash. Keep all unnecessary personnel well clear of ash-handling operations due to the possibility of dust being carried downwind.

(8) Ensure that ash disposal containers are covered.

(9) Beware of overhead discharge.

(10) Only remove the cover when access is necessary and replace the cover as soon as possible.

(11) Do not eat, drink, or smoke while handling ash.

(12) Remove gloves before removing eye protection to avoid ash from gloves falling into eyes.

(13) After handling ash, remove hard hats carefully so that ash does not fall into eyes.

(14) If ash enters eyes, immediately flush eyes.

(15) Wash hands immediately after handling ash and before eating, drinking, or smoking.

Work: Handling Grit

PRACTICE

(1) Use care and caution when working around grit.

(2) When shoveling or lifting grit, wear a back support, gloves and appropriate eye protection.

(3) Ensure that only grit is disposed of in grit disposal containers.

(4) When transporting grit containers with a forklift, follow the organization's Forklift Safety Procedures.

(5) Use care and caution when entering a grit disposal dumpster. Moreover, if trash or screenings have been disposed of in dumpster, watch out for sharp objects.

(6) When climbing into a grit container to spread out grit to make it even, wear rubber boots and position your feet so that you are not shoveling in an odd position.

(7) When entering a building where grit disposal containers are kept, ensure that the ventilation system (i.e., scrubber system) is working properly. If the ventilation system/scrubber system is down, test the atmosphere with an air monitor before entering. Do not enter if the atmosphere is unacceptable (H_2S > 20 ppm, O_2 < 19.5%, or combustibility > 10% LEL).

(8) When entering a grit hopper, ensure that it is lockedout/taggedout using the organization's Lockout/Tagout Procedure.

(9) Inform coworkers/supervisors before entering a disposal dumpster or hopper.

(10) Ensure proper lighting.

(11) Beware of overhead discharge.

(12) When working with classifiers or grit equipment, beware of rotating equipment.

(13) Once grit disposal containers are full, ensure that they are hauled away promptly.

(14) Do not eat, drink, or smoke while handling grit.

(15) Wash hands immediately after handling grit.

(16) Ensure that open wounds are covered and protected while handling grit.

Work: Bicycling

PRACTICE

(1) Only bicycles purchased and approved by the organization may be operated on the plant site while performing organizational business.

(2) At the beginning of each shift, inspect the bicycle you will be using to

ensure that the brakes work, seat is adjusted to the proper size, chains are lubricated, tires have sufficient tread, tires are inflated (to the manufacturer's recommended air pressure – located on side of tire), kickstand is working, the body of the bicycle is in sound shape, and that there are no loose bolts.

Moreover, any mechanical problems with the bicycle are noted in the daily log book and the bicycle is taken out of service until the problem is fixed.

(3) Ride bicycle on the right side of the street.

(4) If there is no ramp to access a sidewalk, avoid riding on it.

(5) Obey all traffic laws, signs, signals, and pavement markings.

(6) Give cars, trucks, forklifts, and pedestrians the right of way.

(7) Avoid broken pavement, litter, potholes, loose gravel or anything that can cause you to lose control of the bicycle.

(8) When in a group, ride single file.

(9) Objects that are more than 20 lbs or larger than the carrier basket are not to be transported on bicycles.

(10) When transporting samples by bicycle, ensure that they are carried in the carrier basket.

(11) Avoid riding bicycles during periods of heavy rainfall, snow, and ice.

(12) Wear hard hats when riding bicycles.

(13) Always look behind you before turning or changing lanes.

Work: Personal Hygiene and Safety

PRACTICE

(1) Practice good personal hygiene and safety to guard against occupationally related diseases.

(2) Wear clothing that protects the arms and legs.

(3) Avoid loose-fitting clothing that could get caught in moving equipment.

(4) Keep shirts tucked into pants.

(5) Change work clothing on a regular basis and more often when clothing becomes extremely dry.

(6) Launder work clothes separately from the regular family wash. After washing work clothes, clean the inside of the washing machine with a disinfectant, such as Lysol®.

(7) Wear steel-toed safety shoes.

(8) Wear the appropriate protective gloves whenever working in contact with wastewater, biosolids, or chemicals. Always check gloves for leaks before starting work. Consult material safety data sheets on chemical(s) to ensure that the correct gloves are worn when handling chemicals.

(9) Wear protective gloves whenever the hands have cuts or broken skin. *Never* allow wastewater, biosolids, or chemicals to come into contact with cuts or broken skin.

(10) Do not place fingers into mouth, nose, ears, or eyes while handling wastewater, biosolids, or chemicals.

(11) Wash hands with a disinfectant soap after handling wastewater, biosolids, or chemicals.

(12) Wash hands with a disinfectant soap before eating, smoking, or going to the lavatory.

(13) Avoid breathing fumes, dust, or vapors. Wear the appropriate respiratory protection.

(14) Do not smoke or eat while working around wastewater treatment tanks and equipment.

(15) Drink potable water only from drinking fountains. *Never* drink water from a hose bib or sill cock.

(16) Shower after each day's work.

(17) Change bandages covering wounds frequently and ensure wounds are kept clean.

(18) Use the correct PPE for each job task.

(19) Do not sacrifice safety for speed.

(20) Do not run on the plant site *except* in an emergency.

(21) Do not play practical jokes or contribute to rowdiness and horseplay on the plant site.

(22) Do not report to work under the influence of alcohol or drugs. Do not use alcohol or drugs while on company time.

(23) Store food and drink only in refrigerators and other spaces so designated.

Work: Office and Clerical Work

PRACTICE

(1) Walk at all times. Use care in passing through doorways.

(2) Reading while walking around the office or corridor is inviting trouble. Keep head up and watch the path of travel.

(3) Use office aisles instead of short-cuts between desks whenever possible.

(4) Check with supervisors before using extension cords. Ensure extension cords are three-prong-type and that cords are not damaged or frayed.

(5) Keep all electrical cords for office machinery and telephones away from office passageways.

(6) Set wastepaper baskets away from passageways so that employees do not stumble over them or step into them.

(7) When carrying bulky objects, be sure vision is not blocked. Try to avoid carrying loads that cause both hands to be encumbered.

(8) Stairways and passageways should be unobstructed by any type of equipment, furniture, tools or other articles. Do not store materials under stairways.

(9) Exercise special care on stairways. Always use handrails.

(10) Slippery floors caused by water or other spilled liquids are dangerous and should be wiped dry immediately. Bottles should not be left on the tops of desks or cabinets where they can be knocked off and liquid spilled.

(11) Report worn, slippery, or broken flooring immediately.

(12) Do not open more than one filing cabinet drawer at a time; opening more than one file cabinet drawer at a time is dangerous and can cause the cabinet to tip over and injure the worker.

(13) Avoid overloading or creating unbalanced lockers. Exercise caution when stacking items on top shelves or lockers.

(14) When lifting items in the office, lift with knees, not the back. *Never* bend over; instead, squat.

(15) *Never* attempt electrical or mechanical repairs of office equipment; leave it to trained service personnel.

(16) Use approved and inspected step ladders or step stools, not office furniture, for reaching high, inaccessible places.

(17) Do not tip chairs forwards or backwards.

(18) Ensure that chairs and all other office furnishings are in good repair before using them.

(19) Report immediately any odors, smoke, excessive heat, or other signs of burning around electrical equipment, cords, or outlets.

(20) Place space heaters away from people and combustible material.

(21) Keep all desk, table, and file drawers and covers of other furniture and equipment closed when not in use.

(22) Use caution in storing knives, scissors, or razor blades. Always use a razor blade with an approved holder. *Never* store loose razors, knives, or pins in desk drawers.

(23) Dispose of razor blades, pins, broken glass, or other pointed objects in separate receptacles from those reserved for wastepaper.

(24) Papercutter blades should always be in the horizontal position when not in use.

(25) Avoid carrying sharp, pointed pencils in pockets; a broken point may become embedded under the skin.

(26) Avoid paper cuts and scratches by picking paper up by the corner edges.

(27) Cigarette smoking is strictly prohibited in all office spaces.

(28) In case of fire, notify supervisors and warn all other workers in the area calmly and without causing panic. Sound the alarm.

(29) Know the location of all fire extinguishers.

(30) Know emergency evacuation route and exits.

(31) Avoid "horseplay."

(32) Ensure that office areas are ergonomically safe. Contact the Safety Division if an ergonomic study has not been completed on your office space or if you change your office arrangement.

(33) Observe all safety precautions in other departmental offices.

Work: Excavation, Trenching, and Shoring

PRACTICE

(1) All excavations more than 4 feet deep must be evaluated by a "competent person," meaning a person who is capable of identifying existing and predictable hazards in the surroundings or working conditions that are unsanitary, hazardous, or dangerous to personnel, and who is authorized to take prompt corrective measures to eliminate such hazards.

(2) The estimated location of utility installations, or any other underground installations that may reasonably be encountered during excavation work, shall be determined prior to opening an excavation.

(3) While the excavation is open, underground installations shall be protected, supported or removed as necessary to safeguard workers.

(4) Structural ramps that are used by personnel and/or equipment as a means of access or egress from excavations must be designed by a "competent person" qualified in structural design, and shall be constructed in accordance with the design.

(5) Always ensure that a stairway, ladder, ramp, or other safe means of egress is located in trench excavations that are 4 feet or more in depth. Do not have more than 25 feet of lateral travel for employees to reach a means of egress.

(6) Wear a warning vest that is made of reflectorized or high-visibility material when exposed to public vehicular traffic.

(7) Keep a safe distance from any vehicle while it is being loaded or unloaded.

(8) Establish a warning system of barricades, hand signals, mechanical signals, or stop logs when mobile equipment is operated adjacent to an excavation, or when equipment is required to approach the edge of an excavation and the operator does not have a clear and direct view of the excavation edge.

(9) Before any worker enters an excavation 4 feet or more in depth where an oxygen deficient atmosphere or a hazardous atmosphere exists, the atmosphere in the excavation *must* be tested with a direct reading instrument.

(10) Provide positive ventilation to a contaminated or oxygen deficient excavation whenever necessary.

(11) Take proper precautions, use explosion-proof tools and lighting when working within and around an excavation if it exceeds the lower explosive limit (LEL).

(12) *Never* work in an excavation in which there is accumulated water or in excavations in which water is accumulating unless precautions have been taken to pump the water out and away from the excavation site.

(13) Where the stability of adjoining walls or other structures is endangered by excavation operations, support systems such as shoring, bracing, or underpinning shall be provided to ensure the stability of such surroundings.

(14) Excavation below the level of the base or footing of any foundation or retaining wall that could reasonably be expected to pose a hazard to employees *shall not* be permitted, except when a registered professional engineer has approved the determination that the structure is

sufficiently removed from the excavation so as to be unaffected by the excavation activity.

(15) *Never* undermine sidewalks, pavements, or appurtenant structures unless a support system or another method of protection is provided.

(16) Provide adequate protection to prevent loose rock or soil from rolling or falling into an excavation. This may be accomplished by scaling to remove loose material; installing protective barricades at intervals as necessary on the face to stop and contain falling material; or by other means that provide equivalent protection.

(17) Keep materials and equipment at least 2 feet from the edge of excavations. If materials or equipment must be brought within 2 feet of the excavation, use retaining devices that will prevent materials or equipment from falling or rolling into excavations.

(18) Daily inspections of excavations, the adjacent areas, and protective systems shall be made by a "competent person" for evidence of a situation that could result in possible cave-ins. An inspection shall be conducted by the "competent person" prior to the start of work, as needed throughout the shift, after *every* rainstorm, or after any other action that may increase the risk of a cave-in.

(19) Evacuate an excavation if the "competent person" finds evidence of a situation that could result in a possible cave-in.

(20) Walkways or bridges with standard guardrails must be provided when employees or equipment are required or permitted to cross over excavations.

(21) All wells, pits, shafts, etc., must be barricaded or covered upon completion of exploration and similar operations. This may be accomplished by backfilling.

(22) All trenches 5 feet or more in depth must be sloped, benched, shored, or shielded according to OSHA Standard 1926.605. The qualified/competent person in charge of the excavation must classify the type of soil at the site as either stable rock, Type A, Type B, or Type C soil.
   - *Stable rock* refers to natural solid mineral matter that can be excavated with vertical sides and remain intact while exposed.
   - *Type A soil* is cohesive with an unconfined compressive strength of 1.5 tons per square foot. Type A soils include clay, silty clay, sandy clay, clay loam, and hardpan.
   - *Type B soil* is cohesive soil with an unconfined compressive strength greater than .5 tons per square foot but less than 1.5 tons per square foot. Type B soils include granular cohesionless soils like angular gravel, silt, silt loam, and sandy loam.
   - *Type C soil* is cohesive soil with an unconfined compressive

strength of .5 tons per square foot or less. Type C soils include granular soils such as gravel, sand, and loamy sand; submerged soil; soil from which water is freely seeping; or submerged rock that is not stable.

(23) Hard hats and safety footwear are required when trenching or working within a trench.

Work: Traffic Control Devices (Construction Sites)

PRACTICE

(1) At any time the traffic flow is disrupted, the appropriate local authorities must be notified to secure the required approval and/or permits.

(2) Only an employee possessing a valid Flagger Certification Card may control traffic for construction and maintenance operations in accordance with State Department of Transportation Regulations.

(3) Flaggers are required to carry their Flagger Certification Card with them. They must also wear an orange vest, hard hat, and safety footwear while flagging.

(4) Employees working in or near a road must wear orange vests, hard hats, and safety footwear.

(5) Traffic control situations and signing setups will vary with local conditions and the requirements of various municipalities. Ask your supervisor and/or refer to the State Work Area Protection Manual.

Work: Traffic

PRACTICE

(1) Provide adequate protection to and from vehicle and pedestrian traffic when working in or adjacent to public and plant roads.

(2) Use traffic signs, cones, and barricades to direct traffic safely around the work site.

(3) Ensure that a certified flagman is directing traffic in congested or highly traveled areas.

(4) Park work vehicles so as to minimize hazards to oncoming traffic.

(5) Park work vehicles between the work site and oncoming traffic whenever possible.

(6) Ensure that all employees working in congested or highly traveled areas wear orange vests so that they can easily be seen by passing motorists.

(7) Observe traffic safety rules and always look for traffic before crossing a street or stepping out from the work site.

Work: Safety Chains

PRACTICE

(1) Protect the top of all ladders and dangerous stairwells with removable safety chains.

(2) Ensure that all safety chains used can handle a minimum impact of 200 pounds.

(3) Ensure that all safety chains used remain rustfree and in good material condition.

(4) Avoid placing weight on safety chains. *Never* lean against safety chains.

(5) Ensure that all safety chains remain taut with no more than 3 inches of droop.

Work: Rotating Equipment

PRACTICE

(1) Use care and caution when working around rotating equipment.

(2) Turn off, lock and tagout at the main circuit breaker. Then try to start *all* mechanical equipment before working on the rotating parts of the equipment.

(3) Install and maintain guards around all rotating shafts, couplings, belt drives, gears, sprockets, pulleys, chains, and other moving parts that are hazardous to personnel. If you need assistance in determining the correct procedure, contact your supervisor.

(4) Do not operate mechanical equipment if guards are removed.

(5) Avoid wearing loose-fitting clothing that could get caught in rotating equipment.

(6) Keep shirts tucked into pants.

(7) If guards need to be adjusted, ensure that equipment is not running or cycling before adjustments are made.

Work: Hand Tools, Power Tools, and Portable Power Equipment

PRACTICE

(1) Use care and caution when using hand tools, power tools, and porta-ble power equipment.

(2) Do not use tools and equipment unless trained and experienced in the proper use and operation.

(3) Use the proper tools and equipment for the required task. *Never* use tools or equipment in a misapplication.

(4) Inspect tools carefully before using them and discard any tool that ap-pears unsafe.

(5) Use care and caution when using tools with sharp points or edges such as saws, knives, chisels, punches, and screwdrivers. Handtools of this type are not to be set down on surfaces where they can be tripped over, stepped on, or bumped.

(6) Use equipment guards and other safety devices at all times when operating tools and equipment. *Never* bypass a safety guard or switch.

(7) Use safety glasses/goggles and face shields as appropriate.

(8) Inspect tools on a regular basis and before each use to ensure that tools and equipment are in good working order.

(9) Keep tools and power equipment clean and in good operating condi-tion. *Never* use broken handtools and power tools.

(10) Replace wornout tools and equipment.

(11) Use only grounded or double-insulated electrical tools.

(12) *Never* use electrical tools in or near water without ground fault inter-rupter circuit. *Never* stand in water when using an electrical tool or equipment.

(13) Have frayed or broken electrical cords repaired or replaced imme-diately.

(14) Shut off gasoline or diesel engines before refueling whenever possi-ble.

(15) Direct exhaust fumes from gasoline or diesel engines away from work areas.

(16) Apply working force away from the body to minimize the chance for injury if the handtool slips.

(17) Ensure tool handles are fitted properly to tools and free of grease and other slippery substances.

(18) Dress cold chisels, hammers, drift pins, and other tools that tend to mushroom at the head. As soon as they begin to mushroom, check a slight radius (approximately 3/16 inch or 4.7 mm) and grind down the edge of the head.

(19) Do not carry sharp edges or pointed tools in clothing pockets.

(20) Do not use defective wrenches, such as open-end and adjustable wrenches with spur jaws or pipe wrenches with dull teeth.

(21) Do not apply hand tools to moving machinery except when designed for the purpose and necessary in the operation.

(22) Do not throw tools and material from one employee to another, or from one location to another. Use a suitable container to raise or lower small equipment or tools between elevations.

(23) Do not place tools on ladders, stairs, balconies, or other elevated places from which they might create a stumbling hazard or become dislodged and fall.

(24) Inspect and test power tools before each use.

(25) When using an electric drill, secure work by using a clamp, jig, or vise. *Never* hold small work in the hand while drilling. Use adequate eye protection and be sure to remove the chuck key or drift before starting to drill.

(26) Wear a face shield, goggles, or face shield *and* safety glasses whenever using a grinding wheel. *Never* grind on the side of a grinding wheel unless it is designed for side grinding.

(27) Prior to installing new abrasive wheels on grinders, perform a "ring" test on wheel. The *ring test* is conducted by striking the wheel with a metal object (small hammer) and listening to hear a sharp metallic ring. If a dull sound is heard instead, the wheel may be defective; replace it with a wheel that passes the ring test.

(28) Run new abrasive wheels for at least one minute before applying work to the wheel surface.

(29) Before starting a pedestal grinder, ensure that the work area is clear and that the wheel rotates freely. After starting the grinder, stand to one side while the wheel reaches speed.

Work: Coating/Painting Operations

PRACTICE

(1) Use care and caution when handling coatings, paints, lacquers, thinners, or solvents. Avoid inhaling the vapors. Wash hands thoroughly

after using coatings, paints, lacquers, thinners, or solvents.

(2) Wear proper respiratory protection when painting. Consult Material Safety Data Sheet (MSDS) to ensure that the correct type of protection is worn. In an oxygen-deficient or potentially oxygen-deficient environment, wear supplied air or air line respiratory protection.

(3) Use mechanical ventilation when spraying area where dangerous quantities of flammable vapors, mists, combustible residues, dusts, or deposits are present and vent exhaust to a safe location. Ensure adequate ventilation while coating and immediately following coating operations to allow vapors to dissipate.

(4) Do not permit welding, open flame, or sparks in areas where combustible or flammable materials are being sprayed.

(5) Observe all "NO SMOKING" signs posted in spraying areas and paint storage rooms.

(6) Use only approved portable explosion-proof lamps in paint spraying areas in which dangerous quantities of flammable vapors, mists, combustibles, residues, dusts, or deposits are present.

(7) Follow all safety precautions stated in the plant's Confined Space Entry Program when painting in a confined space.

(8) Be familiar with the locations of portable fire extinguishers near paint and coating application areas.

(9) Wear approved eye protection when using spraying equipment.

(10) *Never* store paint and gasoline together. Paint must be stored in a designated flammable liquids cabinet or in an approved room containing approved ventilation, free of rags and debris, and with a proper fire extinguisher close by.

(11) Protect skin from paint and solvents by wearing approved skin covering including gloves, clothing, or protective skin creams.

(12) Observe flash points of materials.

(13) Correctly ground paint equipment being used to protect against static electricity.

(14) *Never* point spray gun toward anyone.

Work: High-Noise Areas

PRACTICE

(1) Observe all posted signs stating "Hearing Protection Required" if work time in an area exceeds the stated time limit.

(2) Whenever working in a designated area at or above the allowable time stated on the warning signs, wear hearing protection.

(3) Wear hearing protection as prescribed by audiometric results at all times when working with a tool listed in the plant's Hearing Conservation Program or when walking through an area with high noise levels.

(4) Ensure that hearing protection is worn properly.

(5) Ensure that hands are clean before inserting any type of ear plug into the ear.

Work: Manholes

PRACTICE

(1) Use care and caution when working around manholes.

(2) Do not smoke, light open flames, or produce sparks in the immediate vicinity of open manholes.

(3) If possible, before removing manhole or entrance covers, test the atmosphere inside the manhole by using a remote sampling probe or aspirator. *If the lower explosive limit is above 10% inside the manhole, do not remove the cover.*

(4) Use picks, hooks, or specially designed devices to open manhole covers and heavy hatch covers.

(5) Lift manhole covers and heavy hatches with the legs. *Never* lift with the back muscles.

(6) Lay removed manhole and heavy hatch covers flat on the ground several feet away from the opening.

(7) When manhole or entrance covers are removed, the opening must be promptly guarded by a railing, temporary cover, or other temporary barrier that will prevent an accidental fall through the opening and protect each employee working in the space from foreign objects entering the space.

(8) Use barricades and/or warning devices to direct traffic around open manholes.

(9) Follow the plant's Confined Space Entry Program and all safety procedures before entering and during entry into any manhole.

(10) Have at least two persons present before entering a manhole: One person to enter the manhole and one person in the clear to observe in the event of an emergency. One of the persons *must* be a "qualified person" as defined by the plant's Confined Space Entry Program. This person

bears the responsibility of completing the Confined Space Entry Permit. *Never* enter a manhole unattended.

(11) Wear protective clothing and nonslip, nonsparking shoes in a manhole.

(12) Test each individual manhole step carefully. Be sure to check ladder for overall structural soundness and ability to support weight before using.

(13) Use ladders to access manholes whenever the structural soundness and support ability of the manhole steps/stairs are in question.

(14) Take proper safety precautions when using ladders in a manhole (see the safe work practice on "ladder safety" for precautions).

(15) Use a safety harness and life line when entering a vertical entry manhole, unless the life line creates an entanglement hazard.

(16) Watch out for slippery footing in a manhole.

(17) Do not hand-carry tools and/or equipment while climbing up or down steps or a ladder into a manhole.

(18) Raise or lower tools and/or equipment into a manhole using a rope, sling, or bucket.

(19) Use only explosion-proof portable lighting and nonsparking tools in a manhole.

(20) Avoid using electrical tools in or near water. *Never* stand in water when using electrical tools.

(21) Constant ventilation is required when performing "hotwork" within a manhole.

Work: Lighting

PRACTICE

(1) Adequate lighting is essential for areas used in nighttime operations such as galleries, equipment rooms, and outside areas. If unsure whether adequate lighting exists in an area, contact the safety official.

(2) Adequate lighting is required for structures without windows or below ground level.

(3) Ensure that all walkways, passageways, exits, and stairways are illuminated.

(4) Ensure by testing on a regular basis that standby generators provide power for emergency lighting.

(5) When lighting is required in potentially flammable atmospheres (such as sewers, wetwells, or tanks), use only explosion-proof flashlights and drop lights with heavy duty insulated cords.

(6) By testing on a regular basis ensure that the emergency lights work off battery or other backup power supply.

Work: Lifting, Rigging, and Hoisting

PRACTICE

(1) Use care and caution when lifting heavy tools and equipment.

(2) Inspect loads for size, shape, and weight.

(3) Inspect loads for metal slivers, wooden slivers, jagged edges, burrs, and rough or slippery surfaces.

(4) Clean oil and grease off loads before lifting them.

(5) Clean oil, grease, or other slippery substances off hands before lifting heavy loads.

(6) Wear leather or other approved appropriate gloves when lifting heavy loads. Wear a back support.

(7) Get as close to the load as possible to size it up.

(8) Position your feet far enough apart to obtain balance and stability.

(9) Secure a good footing.

(10) Grip the load firmly.

(11) Keep fingers away from danger points where they could become pinched or crushed.

(12) Bend your legs approximately 90° at the knees.

(13) Keep your back straight.

(14) Lift the load gradually by straightening your legs at the knees while keeping your back straight. *Never* bend over and lift with your back.

(15) Keep the load close to your body.

(16) Do not lift more than can be comfortably handled.

(17) Face the direction in which the load is carried. *Never* twist your body in any direction when lifting a load.

(18) Do not carry a load that blocks your vision.

(19) Do not carry heavy loads up or down stairs.

(20) Summon assistance when a load is determined to be too heavy or bulky for one person to carry.

(21) Lower a load gradually by bending legs at the knees while keeping back straight. *Never* lower load by bending with the back.

(22) Coordinate the joint lifting or lowering of loads with the other personnel involved in the lift.

(23) Balance a load when lifting with two or more people or when using a lifting eye, crane, etc.

(24) When using a lifting device, know the maximum load it is able to safely lift. Ensure that the load to be lifted is less than the maximum load prescribed on the lifting device. *Never* exceed the rated load capacity of lifting equipment.

(25) Attach hooks, hoists, cables, etc., to lifting eyes whenever possible.

(26) Inspect cables, slings, chain fall, and other lifting devices before each use to ensure that they are structurally sound.

(27) Check and recheck cables, slings, etc., used for rigging to ensure that loads are secured.

(28) Lift straight up when using a crane or hoist. *Never* lift with cables at an angle.

(29) Stand clear of suspended loads. *Never* stand or have a portion of the body under a suspended load.

(30) Ensure that all lifting devices are load tested and that proper documentation is maintained.

(31) Ensure that all cranes, trolleys, beams, etc., are marked with the correctly rated load capacity.

Work: Ladders

PRACTICE

(1) Care and caution must be practiced when using ladders.

(2) Always check the Ladder Inspection Tag to ensure that the ladder has been inspected recently. Re-inspect the ladder before use. *Inspect the ladder for*
  - cracked, split, or broken steps, rungs, or braces
  - loose steps or rungs
  - loose metal parts
  - visible wood or metal slivers

(3) *Never* use a broken or damaged ladder. Discard broken or damaged ladders.

(4) Use only ladders that have safety feet.

(5) Use a ladder that fits the length required to do the job. *Never* splice ladders together.

(6) Place ladder feet approximately one-fourth of the ladder height from the top support (see Figure 7.1).

(7) Place ladder feet on solid support.

(8) Place ladder top against a safe support.

(9) Position a ladder so that there is no chance of it slipping or twisting.

(10) Tie off the ladder top to the support so that it cannot slip.

(11) Allow only one person on a ladder at the same time.

(12) Do not use or stand on the top three rungs of a ladder.

(13) Do not lean out from a ladder.

(14) Do not carry tools or equipment while climbing up or down a ladder.

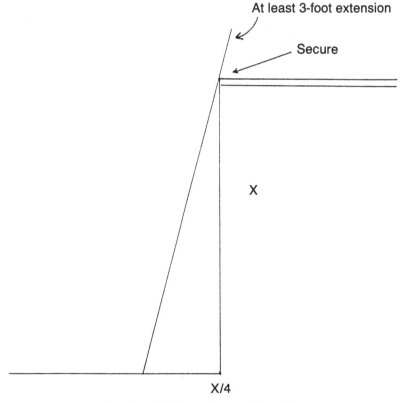

FIGURE 7.1. Proper placement of ladders.

(15) Raise or lower tools or equipment to a person on a ladder using a rope, sling, or bucket.

(16) *Never* stand directly below a ladder.

(17) Spread a step ladder's legs to their fullest extension.

(18) Have a step ladder held by a second person when working 10 feet or more above the floor.

(19) Do not use a step ladder as a straight ladder.

(20) Use only wooden or fiberglass ladders when working around electrical equipment.

(21) Inspect a ladder for structural defects before each use.

(22) Single-piece ladders should extend a minimum of three feet above the top support point.

(23) Never use ladders in a horizontal position as a substitute for a scaffold or a runway between two elevated locations.

Work: Laboratory

PRACTICE

(1) Use care and caution when working in the laboratory.

(2) Store chemicals in a safe place where they are not hazardous to personnel or property.

(3) Label all chemical containers, indicating the chemical name and date of preparation and/or container opening.

(4) Check the labels on chemical containers before using to ensure that the proper chemicals are selected for use.

(5) Properly dispose of unlabeled or out-of-date chemicals. *Never* dispose of chemicals in common trash containers. When hazardous waste is to be disposed, contact laboratory supervisor to ensure that proper disposal procedures are followed.

(6) Read and learn the directions for each chemical's use and safety. This information is found on the chemical's Material Safety Data Sheet (MSDS).

(7) Follow directions carefully. *Never* mix chemicals randomly or indiscriminately.

(8) Handle chemicals carefully when pouring or measuring to prevent spillage.

(9) Immediately clean up chemical spills according to the directions on the chemical's MSDS.

(10) Avoid personal contact with chemicals.

(11) Wear correct gloves for the chemical being handled. Refer to the chemical's MSDS if unsure about the proper hand protection. *Never* touch chemicals with bare hands.

(12) Ensure that protective gloves are free of cracks, tears, and holes and that gloves fit properly before handling chemicals.

(13) Do not place fingers into mouth, nose, ears, or eyes while handling chemicals.

(14) Wash hands with a disinfectant soap after handling chemicals.

(15) Wash off chemical spills on skin immediately with running water.

(16) Do not smoke or eat in the laboratory while handling chemicals or performing tests.

(17) Avoid breathing chemical fumes, dust, or vapors.

(18) Provide positive ventilation to laboratory work areas.

(19) Properly dispose of all broken, chipped, or cracked glassware.

(20) Do not use laboratory glassware as coffee cups or food containers.

(21) Use a suction bulb to pipette chemicals or wastewater. *Never* use mouth to suck up a fluid or chemical in a pipette.

(22) As required by the applicable MSDS, use safety goggles and/or face shield when transferring and measuring chemicals, or whenever there is a potential for chemicals to be splashed in the eyes.

(23) Use tongs or thermal gloves when handling hot utensils.

(24) Use only properly grounded electrical equipment.

(25) Always *add acid to water, not* water to acid.

(26) Use carbon dioxide or dry chemical type fire extinguishers to control laboratory fires.

(27) Ensure that the laboratory is equipped with a properly working emergency eyewash/shower. Laboratory workers should know the exact location of each emergency eyewash/shower. Laboratory workers should anticipate the need to use emergency eyewash/showers and be familiar with quickest route to each eyewash/shower station in close proximity to their work station.

(28) Ensure that prominent signs warning workers against hot areas such as ovens, hot plates, furnaces, water baths, and digestion apparatus are posted in the laboratory.

(29) Inspect acid-neutralizing tanks and basins that service lab sinks regularly and recharge with correct neutralizing agent when needed.

(30) Ensure that emergency phone numbers are posted by all telephones.

Work: Housekeeping

PRACTICE

(1) Conduct routine housecleaning.

(2) Keep all work places including service rooms, storage rooms, passageways, and exits clean and orderly.

(3) Maintain storage places for all tools, equipment, and supplies.

(4) Keep tools and equipment in their proper place when not in use. *Never* leave tools or equipment lying around the plant grounds.

(5) Store racks and poles on mounting brackets. *Never* leave rakes or poles lying on the ground or floor.

(6) Use care and caution when stacking material and supplies. *Never* stack materials or supplies to excessive heights.

(7) Keep storerooms clean and free of rubbish and junk.

(8) Keep forced ventilation systems in good working order and free from air blockages.

(9) Keep floors and stairs as dry as possible. When cleaning main walkway areas, post "CAUTION: WET FLOOR" signs.

(10) Keep floor drains free and unplugged.

(11) Use splash guards and drip pans to keep floors clean and dry.

(12) Hose down spills immediately.

(13) Avoid walking in sludge.

(14) Keep all walk areas free of sludge, slime, caustic, polymer, rags, grit, grease, oil, or other materials that could cause slipping.

(15) Remove ice and snow from walkways and other heavily traveled areas.

(16) Place all rags, grit, grease, or other skimmings in open piles on the ground or concrete deck.

(17) Place trash, debris, and rubbish in proper containers.

(18) Keep access and equipment hatches in place and secured except when in use.

(19) Post "WARNING" signs and *protect* against low obstacles such as beams, pipes, valves, and suspended equipment that could strike heads.

(20) Place "WARNING" signs or barricades around open manholes, hatches, gratings, etc.

(21) Provide adequate inside lighting in shop, equipment, and storeroom areas.

(22) Provide adequate outside lighting in walkway and work areas.

(23) Wash down tank bottoms to remove slippery sludge before working in the area.

(24) *Never* enter a sedimentation tank effluent trough or weir trough alone and unattended. Have at least two persons present before entering a launder or weir trough: one person to enter the sedimentation effluent troughs or weir trough, the other in the clear to observe in the event of an emergency.

(25) Use a safety harness and life line when entering a sedimentation tank effluent trough or weir trough as appropriate.

(26) Post "DO NOT DRINK" signs on all nonpotable water supplies, hose bibs, sill cocks, etc.

(27) Extinguish and properly dispose of cigarettes in designated ashtrays.

(28) Immediately clean up any spills of toxic material.

(29) Leaking containers or spigots are to be remedied immediately. Ensure use of respiratory equipment when necessary.

(30) All solvent-soaked rags or absorbents are to be disposed of in airtight metal receptacles and removed from the workplace daily.

Work: Landscaping Equipment

PRACTICE

(1) Use care and caution when working around landscaping equipment.

(2) Do not use landscaping tools unless trained and experienced in their use and operation.

(3) Before starting power tools, make sure all guards are in place.

(4) Check fuel and oil levels of a four-cycle engine. Do not fill the gasoline tank while an engine is running. Do not smoke during refueling.

(5) Wear either safety glasses or goggles while operating lawn mowers.

(6) Before mowing an area, inspect the area and remove stones, branches, and other foreign objects.

(7) Stand clear of grass discharge chute and keep hands and feet from under blade housing when starting a lawn mower.

(8) When mowing on a steep slope, never mow up and down. Mow aross the face of the slope.

(9) Use extreme caution when pulling a mower and do so only for short distances.

(10) Ensure that other personnel and equipment (such as cars) are kept at a safe distance.

(11) While mowing, pay attention to your work at all times.

(12) Do not leave the engine running while a power tool stands unattended. Shut off push mowers before moving across pavement, gravel, or dirt. On riding mowers, disengage the blade drive when moving across pavement, gravel, or dirt.

(13) Stop the engine and disconnect spark plug wire before working on mower.

(14) Do not dismantle or tie off the "dead man control" on lawn mowers. Do not disable automatic shut-off bars on power mowers.

Work: Garage Safety

PRACTICE

(1) All tools used in equipment maintenance should be closeted or properly hung on hooks to prevent falling.

(2) Clean tools to the extent practical after use.

(3) Remove any oil or grease on the garage floor as soon as possible.

(4) Ensure that all lifting devices (floor jacks, lifts, hoists, beams) used have a rated load capacity visible.

(5) *Never* exceed the rated load capacity of a lifting device.

(6) Ensure that hazardous chemicals are stored properly and that their shelf life is current.

(7) *Never* smoke while handling flammable or combustible chemicals.

(8) Run engines inside for as short a time as possible, unless sufficient measures have been taken to duct the exhaust outside.

(9) Wear appropriate personnel protective equipment for the job. See the sections on welding, portable tools, painting, and chemical handling.

(10) Ensure that all drop lights and extension cords used are not cracked and are properly grounded.

(11) *Never* leave a vehicle unattended while it is running.

(12) When working inside a closed-end truck such as a vac haul, treat it as a confined space entry and refer to the plant's Confined Space Entry program for proper precautions to take.

(13) Wear hearing protection when using loud pneumatic tools or machinery.

(14) Wear safety footwear and protective clothing when working in the garage.

(15) Dispose of all waste oil and hazardous waste in accordance with the plant's Waste Oil and Hazardous Waste Disposal Policy.

(16) Store oxygen and acetylene cylinders in an upright position and secure them to a wall at least 20 feet apart with their protective caps and NFPA labels in place.

(17) At the end of the workday, ensure that the lids to parts-cleaning machines are shut and that all tools, chemicals, and machinery are properly secured. Ensure that greasy or oily rags are stored in designated metal containers with closed lids.

Work: Forklift Operation

PRACTICE

(1) Only plant-certified and licensed personnel are permitted to drive a forklift.

(2) Use care, caution, and seatbelts when operating a forklift.

(3) Complete all daily maintenance (fuel, air, water, hydraulics, transmission, battery, brakes, tires, controls) and visual checks. Fill out/initial Daily Check Sheet prior to operation.

(4) Mount and dismount forklift carefully.

(5) Sit on the seat. Keep arms and legs inside of the cab at all times.

(6) Observe traffic and keep to the right.

(7) Do not allow passengers on the forklift.

(8) Slow down or stop at all blind intersections; yield at all others.

(9) Always observe all traffic rules, load limit warnings, and overhead clearances.

(10) Sound warning device at all cross aisles, exits, elevators, sharp corners, ramps, blind corners and when approaching pedestrians.

(11) Face direction of travel. *Never* back up without looking.

(12) Do not exceed rate capacity. Check unit capacity if attachments are installed.

(13) Keep forks 4–6 inches above the ground when traveling.

(14) Travel with load facing uphill on inclines and downgrades.

(15) Operate in reverse when carrying bulky loads and when line of sight over the load is obstructed.

(16) Travel at speeds that allow for safe stops.

(17) Keep forklift and forks clear of pedestrians.

(18) Position the load evenly on both forks.

(19) Ensure that awkward loads are secured.

(20) *Never* permit anyone to stand or pass under the elevated portion of the mast or attachment.

(21) Lower forks, put forklift in neutral, shut off the forklift, set the brake and, if parked on an incline, block the wheels before leaving the forklift. Park the forklift in authorized areas only.

(22) Refuel or recharge batteries only at safe locations. Fully lower load-engaging controls, neutralize motion controls, set brakes, shut off power, remove key, block wheels on inclines, and *do not smoke.*

(23) Do not operate unit in areas without overhead guard in place.

(24) Do not operate any forklift that is in need of repair, defective, or in any way unsafe.

(25) *Never* raise anyone up in the air while they are standing on the forks. Use an appropriate safety platform.

Work: Fire Control and Prevention

PRACTICE

(1) Call the fire department *immediately* in the event of a potentially serious fire.

(2) Unless immediately dangerous to life and health, attempt to control fires until help arrives.

(3) Provide adequate fire fighting equipment in appropriate locations. If assistance is required in determining adequate equipment or appropriate location, contact the plant safety official.

(4) Provide training and instruction on the types and locations of fire fighting equipment and extinguishers and the proper use of each.

(5) Use the proper type of fire extinguisher for the type of fire to be extinguished.

(6) Control a fire by:
   • cooling it to control the heat
   • smothering it to control the oxygen
   • isolating it to control the fuel

(7) Control a Class A fire (combustible wood, cloth, paper, rubbish, etc.) by cooling with water.

(8) Control a Class B fire (flammable liquids, such as gasoline, oil,

grease, etc.) by smothering with foam, carbon dioxide, or dry chemical type fire extinguisher.

(9) Control a Class C fire (electrical equipment) by smothering the fire with the use of carbon dioxide or dry chemical type figure extinguishers. *Never use foam or water type extinguishers on electrical fires.*

(10) Control a Class D fire (combustible metals such as magnesium, sodium, etc.) by special techniques with the use of super Class D dry powder fire extinguisher.

(11) Maintain good housekeeping practices.

(12) Keep stairs, under stairs, passageways, exits, and fire fighting equipment clear of obstructions.

(13) Store flammable materials only in approved sealed containers in approved flammable storage cabinets.

(14) *Never* store gasoline and paint together.

(15) Avoid an accumulation of flammable materials.

(16) Dispose of oil, oily rags, etc., in covered metal containers.

(17) Observe all "NO SMOKING" signs.

(18) Do not smoke, light open flames, or produce sparks in storage areas or around flammable material unless proper safety precautions are taken.

(19) Ensure that all exits are properly marked. If assistance is needed to determine exits or proper marking of exits, contact the plant safety official.

Work: Hazardous Materials

PRACTICE

(1) Use care and caution when handling hazardous materials/chemicals.

(2) Refer to the MSDS for the hazardous material to be used or worked around.

(3) Comply with all Permissible Exposure Limits (PELs).

(4) Conduct initial exposure monitoring and additional monitoring as needed to confirm compliance with PELs.

(5) Limit access to authorized personnel wherever airborne contaminant concentrations may exceed the PEL. No eating, drinking, or smoking is allowed in these areas.

(6) Use protective clothing and ensure that contaminated clothing is removed or stored in a manner that does not contaminate "clean" areas.

(7) After working with hazardous materials, use designated showers to clean up.

(8) Observe all signs and warnings identifying the hazard or hazardous area.

Work: Compressed Gas Cylinders

PRACTICE

(1) Only accept cylinders approved in interstate commerce for transportation of compressed gases. Do not sign shipping papers for the cylinder until a thorough inspection has been made. For example, for 1-ton chlorine cylinders, an inspection should be made to determine the material condition of the cylinder and that proper hazard markings are in place. Also, with an aspirator bottle with ammonia solution in hand, all valves and fusible plugs must be tested to detect any leaks.

(2) Do not remove or change markings.

(3) Protect cylinders from cuts and abrasions.

(4) Transport cylinders weighing more than 30 pounds with a hand- or motorized truck or overhead hoist.

(5) Do not drop cylinders or allow them to strike each other violently.

(6) Do not use cylinders as supports, rollers, or for any purpose other than that for which they are intended.

(7) Do not tamper with safety devices.

(8) Label cylinders as "EMPTY" when empty. Close valves and replace valve protection caps.

(9) Store cylinders in a safe, dry, well-ventilated space.

(10) Do not store flammable materials in the same area as compressed gases.

(11) Never allow a direct flame or electric arc to contact any part of the cylinder.

(12) Use cylinders in an upright position (except 1-ton cylinders of chlorine and sulfur dioxide) and tie them down to prevent tipping.

(13) Never use a compressed gas cylinder without a pressure regulator or manifold.

(14) Never use oil or grease on oxygen cylinder.

(15) Never substitute oxygen for compressed air.

Work: Chemical Handling

PRACTICE

(1) Use care and caution when handling chemicals.

(2) Store chemicals in a safe place where they are not hazardous to personnel, environment, and property.

(3) Label all small portable chemical containers, indicating the chemical name and date of preparation and/or container opening. Larger nonportable chemical containers need a label indicating the chemical name and the correct NFPA label.

(4) Read labels on chemical containers before using to ensure that the proper chemicals are selected for use.

(5) Properly dispose of unlabeled or out-of-date chemicals.

(6) Contact plant safety official when disposal of chemicals is required. *Never* throw away chemicals in common trash cans.

(7) Read and learn directions given on MSDS for each chemical's use and safety.

(8) Follow directions carefully. *Never* mix chemicals randomly and/or indiscriminately.

(9) Handle chemicals carefully when pouring, measuring, or mixing to prevent spillage.

(10) Clean up chemical spills immediately and refer to the chemical's MSDS if unsure about cleanup procedure.

(11) Avoid personal contact with chemicals.

(12) Wear the appropriate gloves when handling chemicals. Always check gloves for cracks, tears, and leaks before using. See MSDS for the correct type of glove to use with a specific chemical. *Never* touch chemicals with bare hands.

(13) Do not place fingers into the mouth, nose, ear, or eyes while handling chemicals.

(14) Wash hands with a disinfectant soap after handling chemicals.

(15) If chemical spills on skin, wash off immediately with running water.

(16) Avoid breathing chemical fumes, dust, or vapors. See MSDS for the appropriate respiratory protection.

(17) Do not eat, drink, or smoke while handling chemicals.

(18) If required by MSDS, wear appropriate eye protection.

Work: Chemical Handling: Chlorine

PRACTICE

(1) Plant personnel *must* be trained and instructed on the use and handling of chlorine, chlorine equipment, chlorine emergency repair kits, and other chlorine emergency procedures.

(2) Use extreme care and caution when handling chlorine.

(3) Lift chlorine cylinders only with an approved and load-tested device.

(4) Secure chlorine cylinders into position immediately. *Never* leave a cylinder suspended.

(5) Avoid dropping chlorine cylinders.

(6) Avoid banging chlorine cylinders into other objects.

(7) Store chlorine 1-ton cylinders in a cool dry place away from direct sunlight or heating units. Railroad tank cars are direct sunlight compensated.

(8) Store chlorine 1-ton cylinders on their sides only (horizontally).

(9) Do not stack unused or used chlorine cylinders.

(10) Provide positive ventilation to the chlorine storage area and chlorinator room.

(11) *Always* keep chlorine cylinders at ambient temperature. *Never* apply direct flame to a chlorine cylinder.

(12) Use the oldest chlorine cylinder in stock first.

(13) Always keep valve protection hoods in place until the chlorine cylinders are ready for connection.

(14) Except to repair a leak, do not tamper with the fusible plugs on chlorine cylinders.

(15) Wear SCBA whenever changing a chlorine cylinder and have at least one other person with a standby SCBA unit outside the immediate area.

(16) Inspect all threads and surfaces of chlorine cylinder connectors for good condition before making any connections.

(17) Use new lead gaskets each time a chlorine cylinder connection is made.

(18) Use only the specified wrench to operate chlorine cylinder valves.

(19) Open chlorine cylinder valves slowly; no more than one full turn.

(20) Do not hammer, bang, or force chlorine cylinder valves under any circumstances.

(21) Check for chlorine leaks as soon as the chlorine cylinder connection is made. Leaks are checked for by gently expelling ammonia mist from a plastic squeeze bottle filled with approximately 2 ounces of liquid ammonia solution. Do not put liquid ammonia on valves or equipment.

(22) Correct all minor chlorine leaks at the chlorine cylinder connection immediately.

(23) Except for automatic systems, draw chlorine from only one manifolded chlorine cylinder at a time. *Never* simultaneously open two or more chlorine cylinders connected to a common manifold pulling liquid chlorine. Two or more cylinders connected to a common manifold pulling gaseous chlorine is acceptable.

(24) Wear SCBA and chemical protective clothing covering face, arms, and hands before entering an enclosed chlorine area to investigate a chlorine odor or chlorine leak.

(25) Provide positive ventilation to a contaminated chlorine atmosphere before entering whenever possible.

(26) Have at least two personnel present before entering a chlorine atmosphere: One person to enter the chlorine atmosphere, the other to observe in the event of an emergency. *Never* enter a chlorine atmosphere unattended. Remember: OSHA mandates that only fully qualified Level III HAZMAT responders are authorized to aggressively attack a hazardous materials leak such as chlorine.

(27) Use supplied-air breathing equipment when entering a chlorine atmosphere. *Never* use canister-type gas masks when entering a chlorine atmosphere.

(28) Ensure that all supplied-air breathing apparatus has been properly maintained in accordance with the plant's Self-Contained Breathing Apparatus Inspection Guidelines as specified in the plant's Respiratory Protection Program.

(29) Stay upwind from all chlorine leak danger areas unless involved with making repairs. Look to plant wind socks for wind direction.

(30) Contact trained plant personnel to repair chlorine leaks.

(31) Stop leaking chlorine cylinders or leaking chlorine equipment (by closing off valve(s) if possible) prior to attempting repair.

(32) Roll uncontrollable leaking chlorine cylinders so that the chlorine escapes as a gas, not as a liquid.

(33) Connect uncontrollable leaking chlorine cylinders to the chlorination equipment and feed the maximum chlorine feed rate possible.

(34) Keep leaking chlorine cylinders at the plant site. Chlorine cylinders received at the plant site must be inspected for leaks prior to taking delivery from the shipper. *Never* ship a leaking chlorine cylinder back to the supplier after it has been accepted (bill of lading has been signed by plant personnel) from the shipper.

(35) Keep moisture away from a chlorine leak. *Never* put water onto a chlorine leak.

(36) Call the fire department or rescue squad if a person is incapacitated by chlorine.

(37) Administer CPR (use barrier mask if possible) immediately to a person who has been incapacitated by chlorine.

(38) Breathe shallow rather than deep if exposed to chlorine without the appropriate respiratory protection.

(39) Place a person who does not have difficulty breathing and is heavily contaminated with chlorine into a shower. Remove their clothes under the water and flush all body portions that were exposed to chlorine.

(40) Flush eyes contaminated with chlorine with copious quantities of lukewarm running water for at least 15 minutes.

(41) Drink milk if throat is irritated by chlorine.

(42) *Never* store other materials in chlorine cylinder storage areas; substances like acetylene and propane are not compatible with chlorine.

Work: Chemical Handling: Sulfur Dioxide

PRACTICE

(1) Plant personnel *must* be trained and instructed on the use and handling of sulfur dioxide (SO_2), SO_2 equipment, SO_2 emergency repair kits, and other SO_2 emergency procedures.

(2) Use extreme care and caution when handling sulfur dioxide.

(3) Lift sulfur dioxide cylinders only with an approved and load-tested lifting device.

(4) Secure sulfur dioxide cylinders into position immediately. *Never* leave a cylinder suspended.

(5) Avoid dropping sulfur dioxide cylinders.

(6) Avoid banging sulfur dioxide cylinders into other objects.

(7) Store sulfur dioxide in a cool dry place away from direct sunlight or heating units.

(8) Store sulfur dioxide 1-ton cylinders on their sides (horizontally) only and not more than one cylinder high.

(9) Provide positive ventilation to the sulfur dioxide storage area and feed room.

(10) *Always* keep sulfur dioxide cylinders at ambient temperature. *Never* apply direct flame to a sulfur dioxide cylinder.

(11) Use the oldest sulfur dioxide cylinder first.

(12) Always keep valve protection hoods in place until the sulfur dioxide cylinders are ready for connection.

(13) Except to repair leak, do not tamper with fusible plugs on sulfur dioxide cylinders.

(14) Wear SCBA whenever changing out a sulfur dioxide cylinder and have at least one other person with a standby SCBA unit outside the immediate area.

(15) Inspect all threads and surfaces of sulfur dioxide cylinder connectors for good condition before making any connections.

(16) Use new lead gaskets each time a sulfur dioxide cylinder connection is made.

(17) Use only the specified wrench to operate sulfur dioxide cylinder valves.

(18) Open sulfur dioxide cylinder valves no more than one full turn.

(19) Do not hammer, bang, or force sulfur dioxide cylinder valves under any circumstances.

(20) If a strong pungent odor of burning sulfur is detected, a leak exists.

(21) Correct all minor sulfur dioxide leaks at the sulfur dioxide cylinder immediately.

(22) Except for automatic systems, draw sulfur dioxide from only one manifolded sulfur dioxide cylinder at a time. *Never* simultaneously open two or more SO_2 cylinders connected to a common manifold. Two or more cylinders connected to a common manifold pulling gaseous sulfur dioxide is acceptable.

(23) Wear SCBA and protective clothing covering head, arms, and hands before entering an enclosed sulfur dioxide area to investigate a sulfur odor or to repair a sulfur dioxide leak.

(24) Provide positive ventilation to a contaminated sulfur dioxide atmosphere before entering whenever possible.

(25) Have at least two personnel present before entering a sulfur dioxide atmosphere: One person to enter the sulfur dioxide atmosphere, the other to observe in the event of an emergency. *Never* enter a sulfur dioxide atmosphere unattended.

(26) Use supplied air breathing equipment when entering a sulfur dioxide atmosphere. *Never* use canister-type gas masks when entering a sulfur dioxide atmosphere.

(27) Ensure that all supplied-air breathing equipment has been maintained in accordance with the Self-Contained Breathing Apparatus Inspection Guidelines as stated in the plant's Respiratory Protection Program.

(28) Stay away from all sulfur dioxide leak danger areas unless helping to make repairs.

(29) Contact trained treatment plant personnel to repair sulfur dioxide leaks.

(30) Secure leaking sulfur dioxide cylinders or leaking sulfur dioxide equipment prior to attempting repair.

(31) Turn uncontrollable leaking sulfur dioxide cylinders so that the sulfur dioxide escapes as a gas, not as a liquid.

(32) Connect uncontrollable leaking sulfur dioxide cylinders to the dechlorination equipment and feed the maximum sulfur dioxide feed-rate possible.

(33) Keep leaking sulfur dioxide cylinders at the plant site. *Never* ship a leaking sulfur dioxide cylinder back to the supplier after it has been accepted from the manufacturer.

(34) Keep moisture away from a sulfur dioxide leak. *Never* put water into a sulfur dioxide leak; it will oxidize into sulfuric acid.

(35) Call the fire department or rescue squad if a person is incapacitated by sulfur dioxide.

(36) Administer CPR (use barrier mask if available) immediately to a person who has been incapacitated by sulfur dioxide.

(37) Breathe shallow rather than deep if exposed to sulfur dioxide without the appropriate respiratory protection.

(38) Place a person who does not have difficulty breathing and is heavily contaminated with sulfur dioxide into a shower. Remove clothing under running water and wash all portions of their body that were exposed to sulfur dioxide.

(39) Flush eyes contaminated with sulfur dioxide with copious quantities of lukewarm running water for at least 15 minutes.

(40) Drink large quantities of warm salt solution to induce vomiting and to reduce the concentration of sulfur dioxide ingested.

(41) Always inspect incoming cylinders for damage or leaks. Do not accept damaged or leaking cylinders. The transporter is responsible for safely handling cylinder leaks that occur before cylinders are accepted by the plant site.

Work: Chemical Handling: Ferric Chloride

NOTE: Ferric chloride along with alum, ferric sulfate, ferrous sulfate, and lime is used in wastewater treatment for improving plant performance. Namely, ferric chloride is used as a precipitant to enhance the degree of suspend solids (ss) and Biochemical Oxygen Demand (BOD) removal.

PRACTICE

(1) Plant personnel must be trained and instructed on the use and handling of ferric chloride, ferric chloride equipment, and ferric chloride emergency response procedures (refer to MSDS).

(2) Use extreme care and caution when handling ferric chloride.

(3) When handling ferric chloride or servicing the ferric chloride feed system, follow Personal Protective Equipment (PPE) guidelines indicated on the MSDS, which may include: rubber boots, gloves, face shield, splash-proof goggles, and impervious coveralls and jacket. Compatible materials for personal protective equipment include butyl rubber, natural rubber, neoprene, and nitrile rubber.

(4) For airborne concentrations of ferric chloride above MSDS PEL or TLV, wear a self-contained breathing apparatus (SCBA) for respiratory protection.

(5) Ensure that an eyewash/shower or hose with potable water is in the immediate area before working on ferric chloride system or with ferric chloride.

(6) Before putting the ferric chloride system into service, check to ensure that the following valves are *open:*
   • desired feed point valves
   • pump suction valves
   • pump discharge valves
   NOTE: If unsure about the location of the above valves, check process schematic.

(7) Before putting the ferric chloride system into service, check to ensure that the following valves are *closed:*

- feed line valves from storage tanks not in use
- pump isolation valves for chemical feed pumps not in use
- pump discharge valves along feed line not in use from chemical pump to feed point
- calibration cylinder valves

NOTE: If unsure about the location of the above valves, check process schematic.

(8) When putting ferric chloride system into service, remember that only one ferric chloride storage tank can supply ferric chloride to the feed pumps at any given time.

(9) Flush chemical feed system for ferric chloride if it is to be out of service for a prolonged period of time.

(10) When taking the chemical feed system for ferric chloride out of service, remember *never* to close a valve while a pump is running. Severe damage or injury could result.

(11) Periodically inspect ferric chloride chemical feed system for leaks.

(12) Report any problems or leaks of ferric chloride to plant supervisors immediately. *NOTE:* A release of ferric chloride of more than 1,000 pounds (90 gallons) *must* be reported using the plant's Chemical Release Reporting Procedure.

(13) Call the fire department or rescue squad if a person is incapacitated by ferric chloride.

(14) Give CPR immediately to a person who has been incapacitated by ferric chloride and is not breathing. (*Caution:* Administration of mouth-to-mouth resuscitation may expose the first aid provider to chemical within the victim's lungs or vomit—use a barrier mask if readily available.) Summon medical attention as soon as possible.

(15) Place a person who does not have difficulty breathing and is contaminated with ferric chloride into a shower. Remove clothing under running water and wash all portions of body that were exposed to ferric chloride. Get medical attention immediately.

(16) Flush eyes contaminated with ferric chloride *immediately* with running water for at least 15 minutes, occasionally lifting the eyelids.

(17) If ferric chloride is accidently ingested, drink large quantities of water and *induce* vomiting. Get medical attention *immediately.*

Work: Chemical Handling of Sodium Hydroxide

PRACTICE

(1) Plant personnel must be trained and instructed on the use and han-

dling of sodium hydroxide (caustic), sodium hydroxide equipment, and sodium hydroxide emergency response procedures (refer to the Material Safety Data Sheet – MSDS).

(2) Use *extreme* care and caution when handling sodium hydroxide.

(3) When handling sodium hydroxide or servicing the sodium hydroxide feed system, follow Personal Protective Equipment (PPE) guidelines indicated on MSDS, which may include using: rubber boots, face shield, safety goggles, and impervious coveralls and jacket. Compatible materials for personal protective equipment include butyl rubber, natural rubber, neoprene, and nitrile rubber.

(4) For concentrations or mists of sodium hydroxide above the MSDS PEL or TLV, wear fully encapsulating suits and self-contained breathing apparatus.

(5) Ensure that an eyewash/shower or hose with potable water is in the immediate area before working on sodium hydroxide system or with sodium hydroxide.

(6) Before putting the sodium hydroxide system into service, check to ensure that the following valves are *open:*
  • desired feed point valves
  • pump suction valves
  • pump discharge valves
  NOTE: If unsure about the location of the above valves, check process schematic.

(7) Before putting the sodium hydroxide system into service, check to ensure that the following valves are *closed:*
  • feed line valves from storage tanks not in use
  • pump isolation valves for chemical feed pumps not in use
  • pump discharge valves along feed line not in use from chemical pump to feed point
  • calibration cylinder valves
  NOTE: If unsure about the location of the above valves, check process schematic.

(8) When putting the sodium hydroxide system into service, remember that only one storage tank can supply sodium hydroxide to the feed pumps at any given time.

(9) Flush chemical feed system for sodium hydroxide if it is to be out of service for a prolonged period of time.

(10) When taking the chemical feed system for sodium hydroxide out of service, remember *never* to close a valve while a pump is running. Severe damage or injury could result.

(11) Periodically inspect sodium hydroxide chemical feed system for leaks.

(12) Report any problems or leaks of sodium hydroxide to plant supervisors immediately. *NOTE:* Remember when leak/spill of sodium hydroxide exceeds 1,000 pounds (approximately 100 gallons), it must be reported using the plant's Chemical Release Reporting Procedure.

(13) Call the fire department or rescue squad for help if a person is incapacitated by sodium hydroxide.

(14) Give CPR immediately to a person who has been incapacitated by sodium hydroxide and is not breathing. (*Caution:* Administration of mouth-to-mouth resuscitation may expose the first aid provider to chemical within the victim's lungs or vomit—use a barrier mask if readily available.) Summon medical attention as soon as possible.

(15) Place a person who does not have difficulty breathing and is contaminated with sodium hydroxide into a shower. Remove clothing under running water and wash all portions of body that were exposed to sodium hydroxide. Get medical attention *immediately.*

(16) Flush eyes contaminated with sodium hydroxide *immediately* with water for at least 15 minutes, occasionally lifting the eyelids.

(17) If sodium hydroxide is accidently ingested, drink large quantities of water *immediately* to dilute. *Do not induce vomiting.* Get medical attention *immediately.*

Work: Chemical Handling of Muriatic Acid

PRACTICE

(1) Plant personnel must be trained and instructed on the use and handling of muriatic acid, muriatic acid equipment, and muriatic acid Emergency Response Procedures (refer to the Material Safety Data Sheet—MSDS).

(2) Use *extreme* care and caution when handling muriatic acid.

(3) When handling muriatic acid or servicing the muriatic acid feed system, follow Personal Protective Equipment (PPE) guidelines indicated on the MSDS, which may include use of: rubber boots, gloves, face shield, splash-proof safety goggles, and impervious/chemical-resistant coveralls and jacket. Compatible materials for personal protective equipment include butyl rubber, natural rubber, neoprene, nitrile rubber, polyvinyl chloride, viton, saranex, and polycarbonate.

(4) When handling muriatic acid in the laboratory, goggles, face shield, lab coat or apron, gloves, and a vent hood need to be used.

(5) When using/handling muriatic acid in a confined space, a fully encapsulating suit and self-contained breathing apparatus (SCBA) should be worn to prevent contact with high vapor or fume concentrations in air.

(6) Ensure that all storage containers of muriatic acid are labeled. Ensure that the correct NFPA label is on the container (i.e., Blue-3, Red-0, and Yellow-0).

(7) When diluting muriatic acid, *always* add the muriatic acid to water slowly and cautiously. Preferably no faster than it is consumed by the reaction.

(8) Store muriatic acid in a corrosion-proof area; isolate it from incompatible materials, such as metals and *do not* store muriatic acid near oxidizing materials.

(9) Report any problems with or leaks of muriatic acid to plant/work center supervisors immediately. Remember, a release of muriatic acid of more than 5,000 pounds or (500 gallons) must be reported using the plant's Chemical Release Reporting Program (RRP).

(10) For large releases of muriatic acid, Self-Contained Breathing Apparatus (SCBA) and full protective clothing is required for employees to clean up the spill.

(11) When cleaning up muriatic acid, ventilate area and neutralize spill with soda ash or lime. With a clean, plastic shovel, *carefully* place material into a clean, dry container and cover container. Finally, flush spill area clean with water, remembering that washed-away residue is to be considered as hazardous waste and should be disposed of accordingly.

(12) If muriatic acid is accidentally ingested, *get immediate medical help. Do not induce vomiting.* If victim is conscious, give water, milk, or milk of magnesia.

(13) If a person is overcome by muriatic acid, remove victim to fresh air. Get medical attention immediately. If the victim is not breathing, give artificial respiration using a barrier mask to prevent contamination.

(14) If muriatic acid is accidentally spilled on the skin, immediately flush the skin with running water. If muriatic acid is on clothing, remove clothing and wash before clothing is worn again. Get medical attention immediately if skin irritation has occurred.

(15) If muriatic acid gets into the eyes, *immediately* flush eyes with running water for at least 15 minutes. Get medical attention immediately.

Work: Chemical Handling of Anhydrous Ammonia

PRACTICE

(1) Plant personnel must be trained and instructed on the use and handling of anhydrous ammonia, anhydrous ammonia equipment, and anhydrous ammonia emergency response (refer to the Material Safety Data Sheet–MSDS).

(2) Use extreme care and caution when handling anhydrous ammonia.

(3) Do not allow development of excessive pressure in the anhydrous ammonia tank. Excessive pressure generally occurs during tank filling or fire. A pressure relief valve in the center of the tank's manifold will release ammonia gas when there is excessive pressure in the tank.

(4) Wear the Personal Protective Equipment (PPE) indicated on the MSDS when working on the anhydrous ammonia system. This may include rubber boots, gloves, face shield, splash-proof safety goggles, impervious coveralls and jacket, and self-contained breathing apparatus (SCBA). Compatible materials for personal protective equipment include butyl rubber, natural rubber, neoprene, nitrile rubber, and polyvinyl chloride. *NOTE:* Ensure that all supplied-air breathing equipment has been maintained in accordance with the Self-Contained Breathing Apparatus Inspection Guidelines as stated in the plant's Respiratory Protection Program.

(5) Before working on anhydrous ammonia system or with anhydrous ammonia, ensure an eyewash/shower or hose with potable water is in the immediate work area.

(6) Maintain single-phase operation when withdrawing from the tank before and after the evaporators. Whenever the liquid phase changes to the gas phase within the transfer lines, obstruction of the line flow may occur due to freezing.

(7) Ensure the pressure regulator (PRV) on the ammoniators is set at 40 psi. When pressure within the ammoniators is built up greater than 100 psi, excess gas is vented to the atmosphere via a relief valve.

(8) Ensure proper dosage control for ammoniators is set. This is especially important if one ammoniator is down for maintenance, etc.

(9) Stay upwind from all anhydrous ammonia leak danger areas unless helping to make repairs. Look to wind socks for wind direction.

(10) Contact trained plant personnel to repair problems with the anhydrous ammonia system.

(11) Report all leaks in the anhydrous ammonia system to plant supervisors

immediately. *NOTE:* A release of anhydrous ammonia of more than 100 lb *must* be reported using the plant's Chemical Release Reporting Program (RRP).

(12) Breathe shallow rather than deep if exposed to anhydrous ammonia without the appropriate respiratory protection.

(13) Wear SCBA and protective clothing covering the head, arms, and hands before entering an enclosed anhydrous ammonia area to investigate an ammonia odor or leak.

(14) Provide positive ventilation to a contaminated anhydrous ammonia atmosphere before entering whenever possible.

(15) Have at least two personnel present before entering an ammonia atmosphere: One person to enter the ammonia atmosphere, the other to observe in the event of an emergency. *Never* enter an ammonia atmosphere unattended.

(16) Call the fire department or rescue squad if a worker is incapacitated by anhydrous ammonia.

(17) Give CPR immediately to a worker who has been incapacitated by anhydrous ammonia and is not breathing. (*Caution:* Administration of mouth-to-mouth resuscitation may expose the first aid provider to chemical within the victim's lungs or vomit—use barrier mask if readily available.) Summon medical attention as soon as possible.

(18) Place an exposed worker who does not have difficulty breathing and is contaminated with anhydrous ammonia into a shower. Remove clothing under running water and wash all portions of body that were exposed to ammonia. *NOTE:* Forcible removal of clothing frozen to skin may tear skin. If appropriate, thaw clothing before removal. Get medical attention *immediately.*

(19) Flush eyes contaminated with ammonia *immediately* with running water for at least 15 minutes, occasionally lifting the eyelids. Get medical attention as soon as possible.

(20) If anhydrous ammonia is accidentally ingested, drink large quantities of water *immediately* to dilute and *do not induce vomiting.* Get medical attention *immediately.*

Work: Electrical

PRACTICE

(1) Use care and caution when working around *all* electrical equipment.

(2) Observe "DANGER" and "HIGH VOLTAGE" signs.

(3) Stay clear of areas marked as hazardous.

(4) Only *qualified* and *authorized* personnel are to work on electrical equipment. *Qualified Electrical Personnel* is defined as follows:
  - Must have the skill necessary to distinguish live exposed parts from other parts of electrical equipment.
  - Must have the skill necessary to determine the nominal voltage of exposed live parts.
  - Must have the knowledge to determine the clearance distances specified for working with or in the vicinity of various voltage lines.

(5) Follow the plant's lockout/tagout procedure.

(6) Consider all electrical conductors and equipment to be "live" until positively proven to be de-energized.

(7) Do not bypass electrical safety devices.

(8) Ensure that all electrical controls and switches are well marked and accessible.

(9) Use only grounded and double-insulated electrical tools.

(10) When using electrical tools in or near water, use *only* tools that are connected to ground fault interrupters.

(11) Have frayed or broken electrical cords repaired immediately. *Never* tape frayed or damaged electrical cords.

(12) Use only wooden or fiberglass ladders when working around electrical lines and equipment.

(13) Use only explosion-proof portable or intrinsically safe lighting when working in combustible or explosive atmospheres such as manholes, enclosed chambers, digester tanks, etc.

(14) Do not use open-face drop lights.

(15) Use carbon dioxide (C-type) or dry chemical type (B-type) fire extinguisher to control electrical fires. *Never* use foam type fire extinguisher on electrical fires.

(16) *Never* use an empty electrical control panel as a storage locker.

Work: Pulling Electrical Cable

PRACTICE

(1) Only *qualified* and *authorized* electricians or electrical cable pullers are to pull or run electrical cables.

(2) Use care and caution when working around *all* electrical equipment.

(3) Consider all existing electrical wires as "live" until positively proven de-energized.

(4) Follow the plant's lockout/tagout procedure.

(5) Use a properly inspected wood or fiberglass ladder when pulling cables.

(6) Wear safety goggles or glasses meeting the ANSI Z87 Standard while pulling electrical cables.

(7) Wear dry leather gloves when pulling electrical cables.

(8) If suspect asbestos-containing material (ACM) (i.e., insulation or ceiling board) is encountered, contact the plant safety official for proper identification of the material in question.

(9) If heavy fiberglass insulation or dust is present, wear a dust mask.

(10) Electrical cabling being installed needs to be free of breaks, frays, and exposed wiring.

(11) If two or more workers are involved in running or pulling cable, they must have a means of communicating with each other.

(12) Workers need to ensure that dropped ceiling tiles or insulation materials are put back into place and properly secured.

Work: Vehicular and Operating Equipment

PRACTICE

(1) Do not use vehicles or operating equipment unless authorized to do so.

(2) Do not use vehicles or operating equipment unless trained and experienced in their use and operation.

(3) Observe all rules and regulations when operating vehicles or equipment.

(4) Report any ticketed violations occurring in vehicles or operating equipment to management *immediately.*

(5) *Always* wear seat belts and use other safety devices whenever operating vehicles or equipment.

(6) *Do not exceed posted speed limits.*

(7) Horseplay is not permitted while operating vehicles of operating equipment.

(8) Do not sleep in vehicles or operating equipment.

(9) Turn off engine before refueling vehicle or operating equipment.

(10) Turn off engine whenever leaving vehicle or operating equipment.

(11) Set the hand brakes and place a manual transmission in gear whenever parking vehicle or operating equipment.

(12) Report malfunctioning vehicle or operating equipment immediately to management.

(13) Possess a valid operator's license for all vehicles you operate. For example, use of a forklift requires a Forklift Driver's License; operation of very large vehicles may require a Commercial Driver's License (CDL).

(14) *Never* operate a vehicle or other operating equipment while under the influence of drugs or alcohol.

Work: Lead-Based Paint Abatement

PRACTICE

(1) If lead-based paint is suspected, notify the plant safety official.

(2) If the plant safety official confirms lead-based paint, any abatement (removal) procedures must be performed by licensed lead abatement personnel.

Work: Ventilation

PRACTICE

(1) Adequate ventilation is required to support life, to prevent the formation of explosive gas mixtures, and to maintain a safe working environment.

(2) At the beginning of each shift, check all supply air differential gauges to ensure passage of airflow at designed rate.

(3) Check ventilating hoods periodically for air flow.

(4) Use portable blowers when working in manholes, sumps, wetwells, and normally submerged areas that have been drained for access. Any area with potential oxygen deficiency, hazardous, or explosive atmosphere must be thoroughly ventilated prior to entry.

(5) Before starting to work in potentially hazardous atmosphere or confined spaces, follow Confined Space Entry Procedures.

(6) Maintain ventilation equipment in operation while working.

(7) When welding, performing other types of hotwork, or painting, ensure that adequate ventilation is provided.

Work: Drinking (Potable) Water

PRACTICE

(1) *Never* drink water from a hose, hose bib, or sill cock.
(2) Post "Do Not Drink" signs on all nonpotable water supplies, hose bibs, sill cocks, etc.
(3) Eliminate all cross-connections without back flow preventers between potable and nonpotable water supplies.
(4) Insure potable water backflow preventers function and are properly maintained.

Work: Welding and Torch Cutting

PRACTICE

(1) Use care and caution when welding to avoid electrical shock, burns, radiant energy (including "flash"), toxic fumes, fires, and explosions.
(2) Wear approved welding goggles or helmets (hoods) with the proper shaded lens. Wear proper eye protection to guard against flying particles when the helmet is raised.
(3) Use ear protection when welding or cutting overhead.
(4) Wear gauntlet gloves while welding or cutting.
(5) Ensure that outer clothes are free of oil or grease. Fasten clothing around the neck, wrists, and ankles. Wear protective leathers.
(6) Make sure the work area is well ventilated when chlorine solvents are in use, or while brazing, cutting or welding cadmium, beryllium, chromium, zinc, brass, bronze, or galvanized or lead-coated material.
(7) If ventilation is not possible or is inadequate, wear the proper respiratory protective equipment.
(8) Do not weld or cut in dusty or hazardous areas until the areas have been ventilated sufficiently and monitored with a direct reading instrument for combustible gases.
(9) Place safety signs, shields, or barricades around welding jobs for self-protection and for the protection of others from the direct rays of electric arc, welding flame, or splatter.

(10) Prevent hot metal from falling on other individuals, combustible material, or equipment when welding or cutting in elevated positions.

(11) Cool down hot material before leaving the hotwork site.

(12) When welding near energized high-voltage circuits, use solid protective barriers or other means to prevent the ionized air or metallic vapor produced by welding from causing a flashover to the electrical circuit.

(13) Protect flammable material that cannot be removed from the danger of ignition from welding with a shield of noncombustible, fire-resistant material, or take measures to confine the heat, sparks, and slag.

(14) Make sure a suitable fire extinguisher is available during any cutting or welding. In addition, hotwork activities require that an assigned fully trained Fire Watch be present when hotwork is conducted. The Fire Watch must remain at the scene of hotwork activity for 30 minutes after completion of work to make sure that smoldering fires have not been started.

(15) Store gas cylinders upright and chained to secure position. Oxygen cylinders should not be stored within 20 feet of highly combustible materials or cylinders containing flammable gases.

(16) *Never* allow grease or oil to come in contact with gas cylinders, regulators, valves, or connections of gas welding equipment. Do not direct oxygen at oil surfaces or greasy clothes, or into tanks or containers of combustible or flammable liquids.

(17) When making gas welding equipment connections, do not use pipe thread sealant, gaskets, or lubricants.

(18) Check gas connections and fittings with soap or standard testing solutions *before* using gas welding and/or cutting equipment.

(19) Use of a gas torch within an enclosed space requires that the cylinder valves (located outside the enclosed space) be shut off and that the torch and hose be removed from the enclosed space when the torch is not being used.

(20) Use only an approved striker. Do *not* use matches or cigarette lighters to light a torch. Do *not* use heat from hotwork to light a torch.

(21) Install flashguard check valves between hoses and oxygen-acetylene torch to prevent a flashback in the hose.

(22) If a flashback occurs in a torch, shut off the *oxygen* valve first, then close the acetylene valve. Check torch, tip, hose, and gas pressure before relighting the torch.

(23) Do *not* use oxygen or acetylene through a torch or other device with

a shut-off valve unless the cylinder is equipped with a pressure regulator.

(24) Avoid leaving welding electrodes unattended. If electrode holders must be left unattended, remove electrodes and place or protect the holders so they cannot make electrical contact with other individuals or conducting objects.

(25) Turn off equipment when stopping work or leaving the work area.

(26) Keep hoses, cables, and other equipment clear of passageways, ladders, and stairs.

(27) Keep welding equipment, electrode holders, ground clamps, cables, etc., in good working condition. Do not repair gas welding hoses with tape.

(28) Leave valve wrench in place on the valve when a fuel cylinder is in use so that the valve can be turned off quickly in an emergency. To permit quick closing of valves, *do not* open valves on acetylene cylinders more than one and a half turns; acetylene pressure should *never* exceed 15 psig. Open valves on oxygen cylinders all the way.

(29) Welding within a confined space requires constant ventilation, a fire watch, and constant monitoring of the atmosphere with an approved direct reading instrument. Follow *all* precautions as stated in the plant's Confined Space Entry Program.

(30) To avoid a violent reaction, do not use or store acetylene near chlorine.

(31) If welding or torch cutting in a confined space is stopped for some time, special precautions should be taken. For example, the power of arc welding or cutting units should be disconnected and the electrode should be removed from the holder. The torch valves of gas welding or cutting units should be turned off. The gas supply should be shut off at the point outside the confined space. If possible, always remove the torch and hose from the confined space.

(32) Additional information on safe work practices involving welding, cutting, and brazing can be obtained from OSHA's 29 CFR 1910.251–254.

Work: Pump Station Wetwell Entry

PRACTICE

NOTE: Pumping station wetwells should be considered confined spaces unless they (a) are certified safe for entry by a safety professional, (b) have installed ventilation, (c) have more than one way of egress, and (d) have no history of contaminated air or engulfment is not possible.

(1) Entry into pump station wetwells should *always* be made with caution.

(2) Entry should *never* be made until after the interior atmosphere has been certified safe through air sampling using a calibrated, approved air monitor. If the interior atmosphere is *IDLH* (*Immediately Dangerous to Life and Health*), entry is allowed by *Confined Space Permit Only*. Air monitor readings are to be logged in the Pump Station Log.

(3) If the wetwell-installed ventilation system is inoperable, entry is allowed by *Confined Space Permit Only*.

(4) Whenever personnel are within the confines of the wetwell, continuous air sampling *shall* be conducted.

(5) *All* required safety chains in front of and around wet well area *must* be in place.

(6) The two-person rule should *always* apply when wetwell entry is made: One person should remain topside (outside) of wetwell while the other individual is inside. Communication, either visual or verbal, should be maintained between the outside attendant and the entrant.

(7) If wetwell must be entered vertically, it is considered a confined space and can be entered *only* by using Confined Space Entry Procedures.

(8) Confined Space Entry Procedures *must* be utilized if there is any doubt or question about the safety of entering a wetwell.

(9) *Never* smoke in or around a pump station wetwell.

(10) Hotwork permits are required for hotwork activities within a pumping station wetwell.

Work: Fall Protection—General Practice

PRACTICE

(1) Guard all permanent floor openings 12 inches or more in dimension with standard railings.

(2) Guard hatchways and chute openings with a hinged floor cover that is to be guarded when open or with removable railings with toeboards.

(3) Guard skylight floor openings with standard screens or railings.

(4) Guard manhole openings with a standard cover. When the cover is removed and the opening is not guarded, a person *shall* constantly guard the hazard.

(5) Guard all temporary openings or have a person in constant attendance to guard against the hazard.

(6) Protect floor holes for fixed machinery so there is no opening wider than 1 inch.

(7) Guard open-sided floors, platforms, and runways that are 4 feet or more above the floor or ground with standard guardrails and toe-boards.

(8) Provide standard railings for all runways, floors, platforms, walkways, etc., above dangerous equipment (tanks, vats, etc.).

(9) Guard all holes or pits in walking surfaces with a cover.

(10) Paint orange or guard all elevations in walking surfaces that could pose a trip hazard.

(11) Life lines and harnesses are to be worn by employees who are working on elevated roofs.

(12) Life lines and harnesses are to be worn by employees who are working on scaffolding that is 6 feet or more above the ground or floor level.

(13) Fixed ladders with more than a 20-foot drop to the ground or floor level are to be equipped with a safety cage or rail.

Work: Fall Protection—Guardrail Systems

PRACTICE

(1) Each worker on a walking/working surface (horizontal and vertical) with an unprotected side or edge 6 feet or more above a lower level shall be protected from falling by the use of *guardrail systems,* safety net systems, or personal fall arrest systems.

(2) The top edge height of top rails on guardrails must be at least 42 inches (nominal).

(3) The height of the top edge may exceed 42 inches.

(4) Midrails, screens, mesh, intermediate vertical members, or equivalent intermediate structural members shall be installed between the top edge of the guardrail system and the walking/working surface when there is no wall or parapet wall at least 21 inches high.

(5) When midrails are used, they must be installed at a height midway between the top edge of the guardrail system and the walking/working level.

(6) Screens and/or mesh must be used when employees or valuable equipment are below the working level and items could get kicked off to the level. The purpose of the screen or mesh is to protect equipment and employees working below.

(7) When screens and/or mesh are used, they need to extend from the top rail to the walking/working level and along the entire opening between top rail supports.

(8) Intermediate members such as balusters, additional midrails, and architectural panels shall be installed such that there are no openings in the guardrail system that are more than 19 inches wide.

(9) Guardrail systems shall be capable of withstanding, without failure, a force of at least 200 pounds applied within 2 inches of the top edge, in any outward or downward direction, at any point along the top edge.

(10) Midrails, screens, mesh, intermediate vertical members, solid panels, and equivalent structural members shall be capable of withstanding, without failure, a force of at least 150 pounds applied in any downward or outward direction at any point along the midrail or other member.

(11) Guardrail systems shall be so surfaced as to prevent injury to an employee from punctures or lacerations, and to prevent snagging of clothing.

(12) Top rails and midrails shall be at least one-quarter inch nominal diameter or thickness to prevent cuts and lacerations.

(13) If wire rope is used for top rails, it shall be flagged at not more than 6-foot intervals with high-visibility material.

(14) When guardrail systems are used at hoisting area, a chain, gate, or removable guardrail section shall be placed across the access opening between guardrail sections when hoisting operations are not taking place.

(15) When guardrail systems are used at holes, they shall be erected on all unprotected sides or edges of the hole.

(16) When guardrail systems are used around holes used for the passage of material, not more than two sides of the hole shall be surrounded by removable guardrail sections to allow passage of materials. When the hole is not in use, it shall be closed over with a cover, or a guardrail system shall be provided along all unprotected sides or edges.

(17) When guardrail systems are used around holes used as points of access (such as ladderways), they shall be provided with a gate, or be so offset that a person cannot walk directly into the hole.

(18) Ramps or runways erected above 6 feet also require guardrails along each unprotected side or edge.

(19) If manila, plastic, or synthetic rope is used for top rails or midrails, they need to be inspected on a daily basis to ensure that they continue to meet the strength requirements listed above.

(20) Any guardrail system found not to meet the strength requirements due to poor design, rust, chemical damage, or alteration will be barricaded off immediately. Moreover, the area will remain barricaded off until repair/replacement of the guardrail system has taken place.

Work: Fall Protection—Safety Net Systems

PRACTICE

(1) Each employee on a walking/working surface (horizontal and vertical) with an unprotected side or edge which is 6 feet or more above a lower level shall be protected from falling by the use of guardrail systems, *safety net systems,* or personal fall arrest systems.

(2) Safety nets shall be installed as close as possible under the walking/working surface on which employees are working, but in no case more than 30 feet below the work level.

(3) When safety nets are used, the potential fall area from the walking/working surface to the net shall be unobstructed.

(4) When used, safety nets shall extend outward from the outermost projection of the work surface as follows:
   • Up to 5 feet vertical distance from working level to horizontal plane of net – the minimum required horizontal distance of outer edge of net from the edge of the working surface is 8 feet.
   • More than 5 feet up to 10 feet vertical distance from working level to horizontal plane of net – the minimum required horizontal distance of outer edge of the net from the edge of the working surface is 10 feet.
   • More than 10 feet vertical distance from working level to horizontal plane of net – the minimum required horizontal distance of outer edge of net from the edge of the working surface is 13 feet.

(5) Safety nets shall be installed with sufficient clearance under them to prevent contact with the surface or structures below when subjected to an impact force equal to the drop test specified in the safe work practice.

(6) Safety nets and safety net installations shall be drop tested at the job site after initial installation and before being used as a fall protection

system, whenever relocated, after major repair, and at 6-month intervals if left in one place.

(7) All drop tests will be documented by the work center supervisor and documentation of drop tests for safety net systems will remain on file indefinitely.

(8) The drop test for the safety net system shall consist of a 400-pound bag of sand 30 × 2 inches in diameter dropped into the net from the highest walking/working surface at which employees are exposed to fall hazards, but not from less than 42 inches above that level.

(9) If a drop test is impossible, the plant safety official with the assistance of a licensed professional engineer will certify that the net and the net installation are in compliance with OSHA CFR 1926.502 Fall Protection Standard. Moreover, a certification record including the identification of the net and its location, the date it was determined that the identified net was installed correctly, and the signatures of the plant safety official and licensed professional engineer shall be completed.

(10) Certification records for safety nets and net installations shall be available at the job site for inspection.

(11) Safety nets shall be inspected at least once a week for wear, damage, and other deterioration. Moreover, the inspections must be documented by the work center supervisor and inspections records must be available for review.

(12) Safety nets shall be inspected after any occurrence that could affect the integrity of the safety net system.

(13) Any materials, scrap pieces, equipment, and tools that have fallen into the safety net shall be removed as soon as possible and at least before additional work is accomplished.

(14) Safety net systems that are purchased or rented shall meet the requirements of OSHA CFR 1926.502 (Standard for Fall Protection).

(15) The maximum size of each safety net mesh opening shall not exceed 36 square inches, nor be longer than 6 inches on any side. The opening, measured center-to-center of mesh ropes or webbing, shall not be longer than 6 inches. All mesh crossing shall be secured to prevent enlargement of the mesh opening.

(16) Each safety net system shall have a border rope for webbing with a minimum breaking strength of 5,000 pounds.

(17) Connections between safety net panels shall be as strong as integral net components and shall be spaced not more than 6 inches apart.

Work: Fall Protection—Personal Fall Arrest Systems

PRACTICE

(1) Each employee on a walking/working surface (horizontal and vertical) with an unprotected side or edge shall be protected from falling by the use of guardrail systems, safety net systems, or *personal fall arrest systems*.

(2) Life lines shall be protected against being cut or abraded.

(3) Lanyards and vertical life lines shall have a minimum breaking strength of 5,000 pounds.

(4) Snaphooks and D-rings shall have a minimum tensile strength of 5,000 pounds. Snaphooks and D-rings shall be proof-tested to a minimum tensile load of 3,600 pounds without cracking, breaking, or permanent deformation.

(5) Snaphooks used in personal fall arrest systems shall be the locking type.

(6) Never attach snaphooks directly to webbing, rope, or wire snaphook or other connector if attached, to a horizontal life line, or to any object that is incompatibly shaped and/or dimensioned in relation to the snaphook such that unintentional disengagement could occur by the connected object being able to depress the snaphook keeper and release itself.

(7) All connectors shall have a corrosion-resistant finish, and all surfaces and edges shall be smooth to prevent damage to interfacing parts of the system. Connectors shall be drop-forged, pressed, or formed steel or made of equivalent materials.

(8) Each employee shall be attached to a separate life line.

(9) Anchorages used for attachment of personal fall arrest equipment shall be independent of any anchorage being used to support or suspend platforms and shall be capable of supporting at least 5,000 pounds per employee attached, or shall be designed as part of a complete fall arrest system that maintains a safety factor of at least two.

(10) Personal fall arrest systems shall not be attached to guardrail systems, nor shall they be attached to hoists.

(11) Personal fall arrest systems shall be designed to bring an employee to a complete stop and limit maximum deceleration distance an employee travels to 3.5 feet.

(12) The attachment point of the body harness shall be located in the center of the wearer's back, near shoulder level, or above the wearer's head.

(13) Body harnesses and components shall only be used for employee protection, not to hoist materials or equipment.

(14) Personal fall arrest systems and components subjected to impact loading (a fall arrest) shall be immediately removed from service and shall not be used again for employee protection until a qualified vender (manufacturer's representative) has inspected and determined whether the system and components are suitable for reuse.

(15) An Emergency Response Plan needs to be developed by the work center supervisor to assure employees a prompt rescue in the event of a fall or assure employees that they can rescue themselves.

(16) Personal fall arrest systems shall be inspected prior to each use for wear, damage, chemical deterioration, or other deterioration, and defective components shall be removed from service.

(17) Body belts *shall not* be used as a part of a personal fall arrest system.

(18) Self-retracting life lines and lanyards that automatically limit free-fall distance to 2 feet or less shall be capable of sustaining a minimum tensile load of 3,000 pounds applied to the device with the life line or to lanyard in the fully extended position.

(19) Self-retracting life lines and lanyards that do not limit free-fall distance to 2 feet or less, ripstitch lanyards, and tearing and deforming lanyards shall be capable of sustaining a minimum tensile load of 5,000 pounds applied to the device with the life line or lanyard in the fully extended position.

(20) Ropes and straps (webbing) used in lanyards, life lines, and strength components of body harnesses shall be made from synthetic fibers.

Work: Asbestos Containing Material (ACM)—Cement Pipe

PRACTICE

(1) When repairs/modifications that require cutting, sanding, or grinding on cement pipe containing asbestos, EPA-trained asbestos workers/ supervisors are to be called to the work site *immediately.*

(2) Excavation personnel will unearth buried pipe to the point necessary to make repairs/modifications. The immediate work area will then be cleared of personnel as directed by the EPA-trained supervisor.

(3) The on-scene supervisor will direct the EPA-trained workers as required to accomplish the work task.

(4) The work area will be barricaded 20 feet in all directions to prevent unauthorized personnel from entering.

(5) EPA-trained personnel will wear *all* required Personal Protective Equipment (PPE). Required PPE shall include Tyvek totally enclosed suits, 1/2 face respirator equipped with HEPA filters, rubber boots, goggles, gloves, and hard hat.

(6) Supervisor will perform the required air sampling prior to entry.

(7) Air sampling *shall* be conducted using NIOSH 7400 Protocol.

(8) A portable decontamination station will be set up as directed by supervisor.

(9) Workers will enter the restricted area *only* when directed by the supervisor and, using wet methods *only,* will either perform pipe cutting using a rotary cutter assembly or inspect the broken area to be covered with repair saddle device.

(10) After performing the required repair/modification, workers will encapsulate bitter ends and/or fragmented sections.

(11) After encapsulation, the supervisor can authorize entry into restricted area for other personnel.

(12) Broken ACM pipe pieces must be properly disposed of following EPA guidelines.

Work: Disposal of Hypodermic Needles (Sharps)

PRACTICE

(1) Hypodermic needles (sharps) used by plant personnel while on duty for approved medical reasons must be disposed of in approved biohazard containers designated for sharps disposal only.

(2) Biohazard containers can be obtained by contacting the plant safety official.

(3) Biohazard containers will be easily accessible to personnel using them and located as close as feasible to the immediate area where sharps are used or can reasonably be anticipated to be found (e.g., washrooms).

(4) Biohazard containers will be maintained upright throughout use.

(5) Biohazard containers will be replaced routinely and not allowed to overflow.

(6) Biohazard containers may be disposed of in regular trash. However, it is the responsibility of the employee(s) using the container to dispose of it.

(7) Damaged biohazard containers or containers that could leak needles may not be disposed of on the plant site.

(8) Employees are not to open biohazard containers.

Work: Sandblasting

PRACTICE

(1) Use care and caution when handling blasting agents and compressed air.

(2) Ensure that the surface to be sandblasted is lead/asbestos free.

(3) Lay out tarps or poly covering to catch residual blast and debris.

(4) Wear proper respiratory protection when sandblasting. Most sandblasting operations require the use of an air line respirator. Supplied-air respirators used in sandblasting operations should be type CE.

(5) Wear appropriate coveralls, work gloves, and foot protection when sandblasting.

(6) Ensure that both the air compressor line and the supplied air line systems have been tested by the plant's safety official within the last 12 months and meet Grade D air requirements.

(7) Check downwind of sandblasting operations for any equipment that may be impacted by sandblasting operations. Move or cover equipment, as appropriate.

(8) Post warning signs stating that sandblasting is being performed and wear *both* eye and hearing protection.

(9) When using internal combustion engine–powered air compressor for respirator air–breathing supply, ensure that valid calibration of CO (carbon monoxide) monitor is in effect, or calibrate CO monitor as required.

(10) Ensure that an attendant is assigned to watch the carbon monoxide monitor and alert the sandblaster if CO is entrained in the supplied air line.

(11) If carbon monoxide becomes entrained in the supplied air line, *immediately* stop operations and remove the person from airline system.

(12) If the person sandblasting detects any strange odors in his/her supplied airline, operations should stop until the odor can be identified and eliminated.

(13) When sandblasting in confined spaces, follow safety precautions in the plant's Confined Space Entry Program.

(14) Ensure that proper fall protection devices are worn/used when sand-

blasting on an elevated work platform (e.g., scaffolding) and inspect scaffolding before working.

(15) Ensure that the deadman control on the blast nozzle is operational.

(16) Inspect hoses to be certain they are not worn; replace defective hoses as necessary.

(17) Ensure that blast hose couplings are wired together.

(18) Ensure that pressure vessels (blast pots) conform to ASME pressure vessel codes.

(19) Wear goggles and leather work gloves when cleaning up after sandblasting operations.

(20) Carefully remove hard hats and hoods after sandblasting to prevent blast debris from entering eyes.

(21) Flush eyes *immediately* if blast debris enters eyes. If eye irritation persists, report injury to supervisor *immediately.*

Work: Calibration of Air-Monitoring Equipment

PRACTICE

(1) When portable air monitors are used for Confined Space Entry or other hazardous operations, monitors must be properly calibrated before use.

(2) For proper air-monitor calibration method and periodicity, follow manufacturer's recommended guidelines.

Work: Storage Batteries

PRACTICE

(1) Use care and caution when transporting, lifting, or servicing storage batteries.

(2) Wear full eye protection, face shield, rubber gloves, rubber boots, and rubber apron when handling electrolyte.

(3) Use insulated tools and remove rings, watches, and other metal jewelry.

(4) Install batteries in rooms or compartments that are ventilated to provide free circulation of air with a minimum of six air changes per hour.

(5) Do not smoke or introduce open flame or spark into battery storage area.

(6) Do not lay tools or any metal parts on battery or allow tools to fall across battery terminals.

(7) Do not wear nylon coats or clothing. Nylon clothing can create static electricity.

(8) Before charging or discharging batteries, ensure vent caps are installed. Vent cap ventilation ports must be free of debris or obstructions.

(9) When connecting cells, arrange cells so that the positive terminal of one cell connects with the negative terminal of the next throughout the battery. The positive lead of the charging source should connect with the positive terminal of the battery, and the negative lead of the charging source connect with the negative terminal of the battery.

(10) Maintain the proper electrolyte level by adding distilled water as required.

(11) Hydrometers used with lead acid batteries *must never* be used with alkaline batteries.

(12) Keep batteries clean and dry. Use only water and baking soda solution for cleaning the cells.

(13) Call the plant electrical division when repair work is needed on batteries.

(14) Report any type of damage to a battery to a supervisor immediately.

(15) Do not store batteries with flammable materials.

(16) Protect batteries from rain, snow, ice, etc.

(17) Report any electrolyte spills to supervisor immediately.

(18) Consult Material Safety Data Sheet (MSDS) for cleaning/neutralization of spilled electrolyte.

(19) If electrolyte contacts skin, wash off immediately.

(20) If electrolyte contacts eyes, thoroughly flush eyes with copious amounts of water for at least 15 minutes. Get immediate medical attention.

Work: Oil-Burning Heaters (Salamanders)

PRACTICE

(1) Do not operate heater in an area with poor or little ventilation.

(2) Do not operate heater in an area with limited means of egress.

(3) Do not use the heater in areas where flammable substances are

handled/stored or in potentially hazardous atmospheres (e.g., confined spaces).

(4) Do not operate a heater that has been damaged, modified, or otherwise changed from its original condition.

(5) Ensure that either an ABC or CO_2 fire extinguisher is present and accessible before operating the heater.

(6) Only properly trained personnel may operate oil-burning heaters.

(7) Oil-burning heaters may only be operated on a stable, noncombustible surface or floor.

(8) Only #1 or #2 diesel fuel, #1 or #2 fuel oil, or kerosene may be used to fuel the heater. Do not use gasoline or fuel that has been contaminated with water.

(9) Do not allow, under any circumstances, water to enter the heater.

(10) Keep the stack cap on the heater when it is not in use and keep the heater out of the rain.

(11) Clean the bowl and replace the fuel frequently (at least after every 100 hours of use).

(12) Ensure that minimum clearances (8 feet top and 6 feet sides) from normal combustible materials are maintained.

(13) Never allow over 6 inches of flame to appear at the top of the stack.

(14) Avoid touching the heater surface while it is operating and for 30 minutes after shutdown.

(15) *Do not move, handle, or fuel heater while hot or in operation. Before refueling heater, wait at least 30 minutes after shutdown.*

(16) Always use the stand supplied with heater; be sure handles are locked to the cover and are securely bolted.

(17) When carrying the heater by the handles, grasp the elbow to provide stability. Never use the handles for hoisting by crane or other lifting devices.

(18) When heater is operating, never look down into the stack.

(19) Never throw paper cups, food, lunch bags, trash or other foreign material down the stack.

Work: Dry/Powdered Chemical Handling

PRACTICE

(1) Use care and caution when handling chemicals.

(2) Refer to the dry/powdered chemical's Material Safety Data Sheet (MSDS) for storage and handling information.

(3) Ensure that good housekeeping procedures are followed when mixing dry/powdered chemicals.

(4) Follow mixing and handling directions carefully. Never mix dry/powdered chemicals randomly or indiscriminately.

(5) Label all small portable chemical containers, indicating the chemical name and date or preparation and/or container opening. Larger non-portable chemical containers require a label indicating the chemical name and the correct warning label (e.g., NFPA label).

(6) Ensure that a box cutter with a retractable blade is used to open bags of dry/powdered chemicals.

(7) Handle dry/powdered chemicals in a manner that prevents spillage when opening, measuring, and mixing.

(8) Clean up dry/powdered chemical spills immediately and refer to the chemical's Material Safety Data Sheet (MSDS) for proper cleanup procedure.

(9) Avoid personal contact with dry/powdered chemicals.

(10) Wear appropriate gloves when handling dry/powdered chemicals. Ensure gloves are in good condition (i.e., no cracks or tears) and that the correct type of glove is used—refer to the chemical's Material Safety Data Sheet (MSDS).

(11) Wear the appropriate eye and face protection when handling dry/powdered chemicals (i.e., safety goggles and full face shield).

(12) Provide mechanical ventilation to prevent dust concentrations and potential exposures.

(13) Wear appropriate respiratory protection as stated on the chemical's Material Safety Data Sheet (MSDS).

(14) Wash hands immediately after handling dry/powdered chemicals.

(15) Carefully brush or wipe off dry/powdered chemicals that get onto work clothes, eye protection devices, hard hat, and other personal protective devices.

(16) Do not eat, drink or smoke while handling dry/powdered chemicals.

(17) Properly secure all container covers, lids, etc., after transferring dry/powdered chemicals into appropriately labeled containers. *NOTE:* Some dry/powdered chemicals readily absorb humidity; thus, the chemical purity can be affected or unwanted chemical reactions can occur.

(18) If first aid is required after handling dry/powdered chemicals, refer to the chemical's Material Safety Data Sheet (MSDS).

(19) Use proper disposal procedures when disposing of dry/powdered chemicals. Never discard chemicals in common trash receptacles.

Work: Working during Lightning Storms

PRACTICE

(1) If a lightning storm approaches, seek shelter *immediately*.

(2) If caught in an open field, crouch down (do not lie down on the ground). If possible, find a low-lying area, but watch out for flash flooding.

(3) Stay away from metal structures such as flag poles, metal support structures, and antennas.

(4) Stay away from trees.

(5) Stay away from power sources.

(6) Do not use the phone or take a shower.

(7) Avoid using water at a sink.

(8) Stay away from windows. Lightning will go through window panes.

(9) Do not take routine samples or use metal sample poles.

(10) If caught in a vehicle, stay in the vehicle. A vehicle is a good insulator against lightning.

(11) If your hair starts to stand on end, lightning may be approaching; crouch down to the ground immediately.

(12) If possible, take only those samples that are absolutely necessary.

Work: Hotwork—Hotwork Permit Required

IINTRODUCTION

Along with Hotwork Permit requirements for work in confined spaces and other hazardous locations, OSHA's Process Safety Management Standard, CFR 29 1910.119, requires plant workers and outside contractors to employ safe work practices when performing hotwork in or near hazardous materials or processes containing or producing hazardous materials. Hotwork permits must specify the proposed work action and the allowable work period.

HOTWORK DEFINED

Hotwork is defined as the use of oxy-acetylene torches, welding equipment, grinders, cutting, brazing or similar flame-producing or spark-producing operations.

PRACTICE

(1) A Hotwork Permit shall be required for plant and outside contractor personnel for any hotwork performed "in or near" hazardous material/ chemical process and facilities as follows:

- Work *on* tanks, containers, piping feed systems or ancillary equipment containing chemicals or fuels.
- Work within 25 feet of a digester (direct flame only).
- Work in chlorine chemical rooms or on any part of nondiluted chlorine system. *NOTE:* Special precautions must be taken when performing gas welding on chlorine systems. *NEVER use acetylene or propane in the presence of chlorine.*
- Work within 25 feet of any flammable or combustible material with an NFPA fire rating of two or greater.
- Wherever confined space entry testing indicates a hazardous atmosphere.
- Wherever a "Hotwork Permit Required" sign is posted.
- Work where, in the plant supervisor's judgment, ignition/explosion of chemicals could occur from sparks, hot slag, etc. *NOTE:* Hotwork Permits *are not* required when the potential hazard can be removed throughout the duration of work; for example, by disconnecting and flushing lines and air sampling.

(2) Hotwork Permits expire upon completion of each indicated task and at the end of the work day. A new permit must be completed and issued at the beginning of each work day or new assignment.

(3) The Hotwork Permit lists required safe work practices that must be documented in the permit and followed during the specified hotwork operations (see Figure 3.32). Any of the safe work practice items listed in the permit that are not applicable to a particular work operation must be noted in the appropriate comment area.

(4) Hotwork operations require the stationing of a qualified Fire Watch. The Fire Watch must be trained and remain on station for at least 30 minutes after completion of hotwork to guard against reflash. For hot-

work not involving the application of direct flame, Fire Watch can be relieved of Fire Watch duties after a thorough inspection of the hotwork area to ensure that there is no danger of reflash from hot metal chips/ slag residue. The names of the Fire Watch and the outside contractor must be entered on the permit.

This glossary defines many of the terms used on Material Safety Data Sheets (MSDS). It also explains some of the significant terms related to safety and health. The glossary can provide substantial assistance in understanding terms commonly used by safety professionals.

Acid. A hydrogen-containing compound that reacts with water to produce hydrogen. Acids are corrosive and may cause severe burns.

Acute Health Effect. An adverse effect on a human or animal body with severe symptoms developing rapidly and coming quickly to a crisis.

Acute Toxicity. The adverse (acute) effects resulting from a single dose of exposure to a substace.

Aerosols. Liquid droplets or solid particles dispersed in air that are of fine enough particle size (0.01 to 100 microns) to remain dispersed for a period of time.

Air Pollutant. Dust, fumes, mist, smoke, and other particulate matter, vapor, gas, odorous substances, or any combination thereof that is emitted into or otherwise enters the ambient air.

Alkali. A substance capable of combining with hydrogen atoms. Any substance that in water solution is bitter, more or less irritating, or caustic to the skin. Strong alkalies in solution are corrosive to the skin and mucous membranes. They are also called *bases*, and may cause severe burns to the skin.

Anhydrous. Does not contain water.

Asphyxiant. A vapor or gas that can cause unconsciousness or death by suffocation (lack of oxygen). Most simple asphyxiant are harmful to the body only when they reduce oxygen in the air (normally about 21%) to dangerous levels (19.5% or lower).

Attendant. Individual stationed outside one or more permit spaces. Monitors the authorized entrants and performs all attendant's duties as assigned by the employer or Confined Space Permit.

Autoignition Temperature. The minimum temperature at which a material ignites without application of a spark or flame. Along with the flash point, autoignition temperature gives an indication of relative flammability.

Base. Substance that usually liberates OH anions when dissolved in water. Bases react with acids to form salts and water. Bases have a pH greater than 7 in solution. Bases may be corrosive to human tissue. A strong base is called alkaline or caustic.

Blanking or Blinding. The absolute closure of a pipe, line, or duct by the fastening of a solid plate that completely covers the base and is capable of withstanding the maximum pressure of the pipe, line, or duct with no leakage beyond the plate.

Boiling Point. The temperature at which a liquid changes to a vapor state at a given pressure; usually expressed in degrees Fahrenheit at sea level pressure (760 mmHg, or one atmosphere). For mixtures, the initial boiling point or the boiling range may be given.

Caustic. The ability of an alkali to cause burns.

Checklist. Checklists are common tools used by safety professionals to audit facilities and work practices.

Chemical. Any element, chemical compound, or mixture of elements and/or compounds where chemicals are distributed.

Chronic Health Effect. An adverse effect on a human or animal body, with symptoms developing slowly over a long period of time.

Chronic Toxicity. Adverse (chronic) effects resulting from repeated doses of or exposures to a substance over a relatively prolonged period of time.

CO. Carbon monoxide, a colorless, odorless, flammable, and very toxic gas produced by the incomplete combustion of carbon; also a byproduct of many chemical processes.

CO_2. Carbon dioxide, a heavy, colorless gas, produced by the combustion and decomposition of organic substances and as a by-product of many chemical processes. CO_2 will not burn and is relatively nontoxic (although high concentrations, especially in confined spaces, can create hazardous oxygen-deficient environments).

Combustible. A term used by NFPA, DOT, OSHA, and others to classify certain liquids that will burn, on the basis of flash points. Both NFPA and DOT generally define combustible liquid as having a flash point of 100°F. Nonliquid substances, such as wood and paper, are classified as ordinary combustibles by the NFPA.

Concentration. The relative amount of a substance when combined or mixed with other substances. Examples: 3 **ppm** hydrogen sulfide in air, or a 75 **percent** caustic solution.

Confined Space. Any area that has limited openings for entry and exit that would make escape difficult in an emergency, has a lack of ventilation, contains known and potential hazards, and is not normally intended or designed for continuous human occupancy.

Corrosive. As defined by DOT, a corrosive material is a liquid or solid that causes visible destruction or irreversible changes in human tissue at the site of contact on — in the case of leakage from its packaging — a liquid that has a severe corrosion rate on steel.

Decomposition. Breakdown of a material or substance (by heat, chemical reaction, electrolysis, decay, or other processes) into parts, elements, or simpler compounds.

Dermal Toxicity. Adverse effects resulting from a material's absorption through skin.

Diffusion Rate. The rate at which one gas or vapor disperses into or mixes with another gas or vapor. The rate is a function of the density of vapor or gas when compared to air (air = 1).

Dilution Ventilation. Air flow designed to dilute contaminants to acceptable levels.

Dispersion. Particulate matter suspended in air or other fluid; the dilution and mixing of contaminants in the ambient environment.

Dose. The term used to express the amount of energy or substance absorbed in a unit volume of an organ. Individual dose rate is the dose delivered per unit of time.

Emergency First Aid Procedures. Actions that should be taken at the time of a chemical exposure or other injury before trained medical personnel arrive.

EPA. U.S. Environmental Protection Agency; federal agency with environmental protection regulatory and enforcement authority.

Evaporation. The process of changing a liquid to a vapor, with mixing into the air or into other gases.

Evaporation Rate. A number showing how fast a liquid will evaporate. The higher the evaporation rate, the greater the risk of vapors collecting in the workplace.

Explosive-Proof Equipment. Apparatus enclosed in a case capable of withstanding an explosion of a specified gas or vapor and preventing the ignition of a specified gas or vapor surrounding the enclosure by sparks, flash, or explosion of the gas or vapor within, and operating at an external temperature so that a surrounding flammable atmosphere will not be ignited.

Exposure. Subjection of a person to a toxic substance or harmful physical agent in the course of employment through any route of entry (e.g., inhalation, ingestion, skin contact, or absorption). Includes past exposure and potential (e.g., accidental or possible) exposure, but does not include situations where the employer can demonstrate that the toxic substance or harmful physical agent is not used, handled, stored, generated, or present in the workplace in any manner different from typical nonoccupational situations.

Exposure to a substance or agent may or may not be an actual health hazard. And industrial hygienist evaluates exposures and determines if permissible exposure levels are exceeded.

Flammability Limits. The range of gas or vapor amounts in the air that will burn or explode if a flame or other ignition source is present.

Flammable. A "Flammable Liquid" is defined by NFPA and DOT as a liquid with a flash point below 100 degrees F.

Flammable Limits. Flammable liquids produce (by evaporation) a minimum and maximum concentration of flammable gases in air that will support combustion. The lowest concentration is known as the lower flammable limit (LFL). The highest concentration is known as the upper flammable limit (UFL).

Flash Point. The lowest temperature at which vapors above a liquid will ignite. There are several flash point test methods, and flash points may vary for the same material depending on the method used.

Fume. Gas-like emanation containing minute solid particles arising from the heating of a solid body such as lead. This physical change is often accompanied by a chemical reaction, such as oxidation. Odorous gases and vapors are not considered fumes.

Gas. A state of matter in which the material has very low density and viscosity, can expand and contract greatly in response to changes in temperature and pressure, easily diffuses into other gases, and readily and uniformly distributes itself throughout the container.

General Exhaust. A system for exhausting air containing contaminants from a general work area.

Gravity, Specific. The ratio of the mass of a unit volume of a substance to the mass of the same volume of a standard substance at a standard temperature. For liquids, water at 4°C is the usual standard. For gases, dry air, at the same temperature and pressure as the gas, is usually the standard.

Hazardous Materials. In a broad sense, a hazardous material is any substance or mixture of substances having properties capable of producing adverse effects on the health and safety of a human being.

Hazardous Waste. Any material listed as such in Title 40 CFR 261, Subpart D, or that possesses any of the hazard characteristic of corrosivity, ignitability, reactivity, or toxicity as defined in Title 40 CFR 261, Subpart C, or that is contaminated by or mixed with any of the previously mentioned materials.

Hotwork. Work involving electric or gas welding, cutting, brazing, or similar flame or spark-producing operations.

Hotwork Permit. The employer's written authorization to perform operations that could provide a source of ignition, such as welding, cutting, burning, or heating.

IDLH, Immediately Dangerous to Life and Health. The maximum concentration from which one could escape within 30 minutes without any escape-impairing symptoms or irreversible health effects. Usually used to describe a condition where self-contained breathing apparatus must be used.

Ignitability. A characteristic that identifies a material as hazardous, when the material has a flashpoint of less than 140°F.

Incompatible. Materials that could cause dangerous reactions from direct contact with one another are described as incompatible.

Ingestion. The taking of a substance through the mouth.

Inhalation. The breathing in of a substance in the form of a gas, vapor, fume, mist, or dust.

Local Exhaust. A system for capturing and exhausting contaminants from the air at the point where the contaminants are produced (welding, grinding, sanding, dispersion operations).

Lockout/Tagout. The placement of a lockout/tagout device on an energy-isolating device, in accordance with an established procedure, to ensure that the energy-isolating device and equipment being controlled cannot be operated until the lockout/tagout device is removed.

Melting Point. The temperature above which a solid changes to a liquid upon heating.

Mist. Suspended liquid droplets generated by condensation from the gaseous to the liquid state, or by breaking up a liquid into a dispersed state, by splashing, foaming, or atomizing.

MSDS. Material Safety Data Sheet. Any MSDS must include information regarding the specific identity of hazardous chemicals. Also includes information on health effects, first aid, chemical and physical properties, emergency phone numbers, and much more.

NIOSH. National Institute for Occupational Safety and Health of the Public Health Service, U.S. Department of Health and Human Services. The agency functions to test and certify respiratory protective devices, recommends occupational exposure limits for various substances, and assists in occupational safety and health investigations and research.

Nonflammable. Not easily ignited, or if ignited, burns very slowly.

OSHA. Occupational Safety and Health Administration of the U.S. Department of Labor. This agency is the safety and health regulatory and enforcement authority for most U.S. industry and business.

Oxidizer. A chemical, other than a blasting agent or explosive, that initiates or promotes combustion in other materials, causing fire either by itself or through the release of oxygen or other gases.

Particulate. The particle of solid or liquid matter. Particulate may be organic or inorganic, and may either be visible or invisible.

PEL. Permissible Exposure Limit; an exposure established by OSHA regulatory authority. May be a Time-Weighted Average (TWA) limit or a maximum concentration exposure limit.

pH. The symbol relating the hydrogen ion concentration of a solution. A pH of 7 is neutral. Numbers increasing from 7 to 14 indicate greater alkalinity. Numbers decreasing from 7 to 0 indicate greater acidity.

ppm. Parts per million; a unit for measuring the concentration of a gas or vapor in air—parts (by volume) of the gas or vapor in a million pars of air. Also used to indicate the concentration of a particular substance in a liquid or solid.

Physical Hazard. In the general safety sense, a hazard of physical origin, such as a fall or a heat burn; not a chemical or infective disease hazard.

PQS. Personnel Qualification Standard; devised by the US Navy, and modified by Robert Rutherford and the author for wastewater treatment. A training outline to be completed by all new employees assigned to a wastewater treatment facility. PQS is to be completed by the new employee before he/she is allowed to begin routine work.

PSM. The Process Safety Management Standard, 29 CFR 1910.119, promulgated by OSHA, effective May 26, 1992.

Reactivity. (Degree of stability.) A description of the tendency of a substance to undergo chemical reaction with itself or other materials with the release of energy.

Respiratory System. The breathing system; includes the lungs and air passages to the air outside the body.

Spill or Leak Procedures. Steps that should be taken if a chemical spill or leak occurs.

TLV. (Threshold Limit Value). An exposure level under which most people can work consistently for 8 hours a day, with no harmful effects.

TQ. (Threshold Quantity in pounds). The amount necessary to be covered by the Process Safety Management standard.

TWA. (Time-Weighted Average exposure). The airborne concentration of a material to which a person is exposed, averaged over the total exposure time.

Unstable. Tending toward decomposition or other unwanted chemical change during normal handling or storage.

Vapor Density. The weight of a vapor or gas compared to the weight of an equal volume of air; an expression of the density of the vapor or gas. Materials lighter than air have vapor densities less than 1.0. Materials heavier than air have vapor densities greater than 1.0.

Vapor Pressure. The pressure exerted by a saturated vapor above its own liquid in a closed container. The higher the vapor pressure, the easier it is for a liquid to evaporate and fill the work area with vapors that can cause health and fire hazards.

Adams, C. E. (1991). Industrial Wastewater Treatment Options, *National Environmental Journal*, Sept./Oct.:16–18.

ANSI. (1989). *American National Standard Requirements for Confined Spaces*, New York: ANSI Z117.1.

CoVan, J. (1995). *Safety Engineering*. New York: John Wiley & Sons.

Department of Health & Human Service. *Guide to Industrial Respiratory Protection.* NIOSH Publication No. 87–116.

Grimaldi, J. V. & Simonds, R. H. (1989). *Safety Management*. Homewood, IL: Irwin.

Kuhlman, R. L. (1989). Notes & Comments: Get off on the Right Foot. *Professional Safety*, March:41.

LeBar, G. (1990). OSHA Plans Update of PPE Rules. *Occupational Hazards*, June:51–53.

Minter, S. G. (1990). A New Perspective on Head Protection. *Occupational Hazards*, June:45–49.

National Safety Council. *Accident Facts, 1989, 91, & 93 Editions.* Chicago, IL.

OSHA. (1987). *Excavation Work Proposal,* April 15.

OSHA. (1993). *Permit-Required Confined Spaces (Permit Spaces).* U.S. Code of Federal Regulations, Title 29, 29 CFR 1910.146.

Rutherford, R. & Spellman, F. (1993). *Personnel Qualification Standard (POS) for Waste-water Operators.* Virginia Beach, VA: Hampton Roads Sanitation District.

U.S. Department of Labor, Bureau of Labor Statistics. (1980). *Accidents Involving Head In-juries.* Report 605. Washington, DC: U.S. Govt. Printing Office.

U.S. Department of Labor, Occupational Safety and Health Administration. (1991). *All about OSHA.* Publication No. 2056. Washington, DC.

U.S. Department of Labor, Occupational Safety and Health Administration. (1989). *Controlling Electrical Hazards.* Publication No. 3075. Washington, DC.

U.S. Department of Labor, Occupational Safety and Health Administration. (1989). *Electrical Standards for Construction.* Publication No. 3097. Washington, DC.

U.S. Department of Labor, Occupational Safety and Health Administration. (1991). *Ergonomics: The Study of Work.* Publication No. 3125. Washington, DC.

219

U.S. Department of Labor, Occupational Safety and Health Administration. (1992). *Hand and Power Tools.* Publication No. 3080. Washington, DC.

U.S. Department of Labor, Occupational Safety and Health Administration. (1990). *Hazard Communication Guidelines for Compliance.* Publication No. 3111. Washington, DC.

U.S. Department of Labor, Occupational Safety and Health Administration. (1989). *Hazardous Waste & Emergency Response.* Publication No. 3114. Washington, DC.

U.S. Department of Labor, Occupational Safety and Health Administration. (1993). *Lead in Construction.* Publication No. 3142. Washington, DC.

U.S. Department of Labor, Occupational Safety and Health Administration. (1995). *Occupational Safety and Health Standards for General Industry—29 CFR Part 1910.* Washington, DC.

U.S. Department of Labor, Occupational Safety and Health Administration. (1992). *OSHA Inspections.* Publication No. 2098. Washington, DC.

U.S. Department of Labor, Occupational Safety and Health Administration. (1988). *Sling Safety.* Publication No. 3072. Washington, DC.

U.S. Department of Labor, Occupational Safety and Health Administration. (1991). *Stairways and Ladders.* Publication No. 3124. Washington, DC.